Lecture Notes in Economics and Mathematical Systems 565

Wolfgang Lemke

Term Structure Modeling and Estimation in a State Space Framework

Springer

Author

Wolfgang Lemke
Deutsche Bundesbank
Zentralbereich Volkswirtschaft / Economics Department
Wilhelm-Epstein-Straße 14
D-60431 Frankfurt am Main
E-mail: wolfgang.lemke@bundesbank.de

ISSN 0075-8442
ISBN-10 3-540-28342-0 Springer Berlin Heidelberg New York
ISBN-13 978-3-540-28342-3 Springer Berlin Heidelberg New York

Springer is a part of Springer Science+Business Media

springeronline.com

Printed in Germany

Typesetting: Camera ready by author
Cover design: *Erich Kirchner*, Heidelberg

Printed on acid-free paper 42/3130Jö 5 4 3 2 1 0

Preface

This book has been prepared during my work as a research assistant at the Institute for Statistics and Econometrics of the Economics Department at the University of Bielefeld, Germany. It was accepted as a Ph.D. thesis titled "Term Structure Modeling and Estimation in a State Space Framework" at the Department of Economics of the University of Bielefeld in November 2004.

It is a pleasure for me to thank all those people who have been helpful in one way or another during the completion of this work.

First of all, I would like to express my gratitude to my advisor Professor Joachim Frohn, not only for his guidance and advice throughout the completion of my thesis but also for letting me have four very enjoyable years teaching and researching at the Institute for Statistics and Econometrics. I am also grateful to my second advisor Professor Willi Semmler. The project I worked on in one of his seminars in 1999 can really be seen as a starting point for my research on state space models. I thank Professor Thomas Braun for joining the committee for my oral examination.

Many thanks go to my dear colleagues Dr. Andreas Handl and Dr. Pu Chen for fruitful and encouraging discussions and for providing a very pleasant working environment in the time I collaborated with them. I am also grateful to my friends Dr. Christoph Wöster and Dr. Andreas Szczutkowski for many valuable comments on the theoretical part of my thesis and for sharing their knowledge in finance and economic theory with me. Thanks to Steven Shemeld for checking my English in the final draft of this book.

Last but not least, my gratitude goes to my mother and to my girlfriend Simone. I appreciated their support and encouragement throughout the entire four years of working on this project.

Frankfurt am Main, August 2005 *Wolfgang Lemke*

Contents

1

Introduction

The term structure of interest rates is a subject of interest in the fields of macroeconomics and finance alike. Learning about the nature of bond yield dynamics and its driving forces is important in different areas such as monetary policy, derivative pricing and forecasting. This book deals with dynamic arbitrage-free term structure models treating both their theoretical specification and their estimation. Most of the material is presented within a discrete-time framework, but continuous-time models are also discussed.

Nearly all of the models considered in this book are from the affine class. The term 'affine' is due to the fact that for this family of models, bond yields are affine functions of a limited number of factors. An affine model gives a full description of the dynamics of the term structure of interest rates. For any given realization of the factor vector, the model enables to compute bond yields for the whole spectrum of maturities. In this sense the model determines the 'cross-section' of interest rates at any point in time. Concerning the time series dimension, the dynamic properties of yields are inherited from the dynamics of the factor process. For any set of maturities, the model guarantees that the corresponding family of bond price processes does not allow for arbitrage opportunities.

The book gives insights into the derivation of the models and discusses their properties. Moreover, it is shown how theoretical term structure models can be cast into the statistical state space form which provides a convenient framework for conducting statistical inference. Estimation techniques and approaches to model evaluation are presented, and their application is illustrated in an empirical study for US data.

Special emphasis is put on a particular sub-family of the affine class in which the innovations of the factors driving the term structure have a Gaussian mixture distribution. Purely Gaussian affine models have the property that yields of all maturities and their first differences are normally distributed. However, there is strong evidence in the data that yields and yield changes exhibit non-normality. In particular, yield changes show high excess kurtosis that tends to decrease with time to maturity. Unlike purely Gaussian models, the mixture models discussed in this book allow for a variety of shapes for the distribution of bond yields. Moreover, we provide an algorithm that is especially suited for the estimation of these particular models.

The book is divided into three parts. In the first part (chapters 2 - 4), dynamic multifactor term structure models are developed and analyzed. The second part (chapters 5 - 7) deals with different variants of the statistical state space model. In the third part (chapters 8 - 9) we show how the state space framework can be used for estimating term structure models, and we conduct an empirical study.

Chapter 2 contains notation and definitions concerning the bond market. Based on a data set of US treasury yields, we also document some stylized facts. Chapter 3 covers discrete-time term structure models. First, the concept of pricing using a stochastic discount factor is discussed. After the analysis of one-factor models, the class of affine multifactor Gaussian mixture (AMGM) models is introduced. A canonical representation is proposed and the implied properties of bond yields are analyzed. Chapter 4 is an introduction to continuous-time models. The principle of pricing using an equivalent martingale measure is applied.

The material on state space models presented in chapters 5 - 7 will be needed in the third part that deals with the estimation of term structure models in a state space framework. However, the second part of the book can also be read as a stand-alone treatment of selected topics in the analysis of state space models. Chapter 5 presents the linear Gaussian state space model. The problems of filtering, prediction, smoothing and parameter estimation are introduced, followed by a description of the Kalman filter. Inference in nonlinear and non-Gaussian models is briefly discussed. Chapter 6 introduces the linear state space model for which the state innovation is distributed as a Gaussian mixture. We anticipate that this particular state space form is the suitable framework for estimating the term structure models from the AMGM class described above. For the mixture state space model we discuss the exact algorithm for filtering and parameter estimation. However, this algorithm is not useful in practice: it generates mixtures of normals that are characterized by an exponentially growing number of components. Therefore, we propose an approximate filter that circumvents this problem. The algorithm is referred to as the approximate mixture filter of degree k, abbreviated by AMF(k). In order to explore its properties, we conduct a series of Monte Carlo simulations in chapter 7. We assess the quality of the filter with respect to filtering, prediction and parameter estimation.

Part 3 brings together the theoretical world from part 1 and the statistical framework from part 2. Chapter 8 describes how to cast a theoretical term structure model into state space form and discusses the problems of estimation and diagnostics checking. Chapter 9 contains an empirical application based on the data set of US treasury yields introduced in chapter 2. We estimate a Gaussian two-factor model, a Gaussian three-factor model, and a two-factor model that contains a Gaussian mixture distribution. For the first two models, maximum likelihood estimation based on the Kalman filter is the optimal approach. For the third model, we employ the AMF(k) algorithm. Within the discussion of results, emphasis is put on the additional benefits that can be obtained from using a mixture model as opposed to a pure Gaussian model.

Chapter 10 summarizes the results. The appendix contains mathematical proofs and algebraic derivations.

2

The Term Structure of Interest Rates

2.1 Notation and Basic Interest Rate Relationships

In this section we introduce a couple of important definitions and relationships concerning the bond market.[1] We start our introduction with the description of the zero coupon bond and related interest rates. A *zero coupon bond* (or a zero bond for short) is a security that pays one unit of account to the holder at maturity date T. Before maturity no payment is made to the holder. The price of the bond at $t \leq T$ will be denoted by $P(t,T)$. For short we will call such a bond a T-bond. After time T the price of the T-bond is undefined. Unless explicitly stated otherwise, we assume throughout the whole book that bonds are default-free.

For the T-bond at time t, $n := T-t$ is called the *time to maturity.*[2] Instead of $P(t,T)$ we may also write $P(t,t+n)$. For the price of the T-bond at time t we sometimes use a notation where the time and maturity argument are given as subscript and superscript, that is we write P_t^n instead of $P(t,t+n)$.

Closely related to the price is the *(continuously compounded) yield* $y(t,T)$ of the T-bond. This is also referred to as the *continuously compounded spot rate.* It is defined as the constant growth rate under which the price reaches one at maturity, i.e. with $n = T - t$,

$$P(t,T) \cdot \exp[n \cdot y(t,T)] = 1 \tag{2.1}$$

or

$$y(t,T) = -\frac{\ln P(t,T)}{n}. \tag{2.2}$$

Again, we will frequently use the alternative notation y_t^n instead of $y(t,t+n)$.

---

[1] For these definitions see, e.g., [65], [94] or [19]. It is frequently remarked in the literature that difficulty arises from the confusing variety of notation and terminology, see, e.g., [63], p. 387.

[2] As in the literature we will also use the word 'maturity' instead of 'time to maturity' when the meaning is clear from the context.

If time is measured in years, then n is a multiple of one year. For instance, the time span of one month would correspond to $n = \frac{1}{12}$. With respect to this convention, the yield defined in (2.2) would be referred to as the annual yield. We also define monthly yields, since those will be the key variables in chapter 3 and 9. If y is an annual yield, then the corresponding monthly yield is given by $y/12$. This can be seen as follows. If time is measured in months, then one month corresponds to $n = 1$. Let for the moment n_M denote a time span measured in months and n_A a time span measured in years. The annual yield satisfies

$$P(t,T) \cdot \exp[n_A \cdot y(t,T)] = 1 \tag{2.3}$$

or

$$P(t,T) \cdot \exp[n_M \cdot \frac{1}{12} y(t,T)] = 1. \tag{2.4}$$

Hence, defining monthly yields as one twelfth of annual yields implies that equation (2.1) is also valid for monthly yields when n denotes the time span $T - t$ in months.

The *instantaneous short rate* r_t is the limit of the yield of a bond with time to maturity converging to zero:

$$r_t := \lim_{n \to 0} y(t, t+n) = -\frac{\partial}{\partial T} \ln P(t,t). \tag{2.5}$$

The *forward rate* $f(t,S,T)$ is the interest rate contracted at t for the period from S to T with $t < S < T$. To see what this rate must be, consider the following trading project at time t. One sells an S-bond and uses the receipts $P(t,S)$ for buying $P(t,S)/P(t,T)$ units of the T-bond. This delivers a net payoff of 0 at time t, of -1 at time S and of $P(t,S)/P(t,T)$ at time T. The strategy implies a deterministic rate of return from S to T. If the forward rate were to deviate from this rate, one could easily establish an arbitrage strategy. Thus, we define the forward rate via the condition $1 \cdot \exp[(T-S)f(t,S,T)] = P(t,S)/P(t,T)$, yielding

$$f(t,S,T) = -\frac{\ln P(t,T) - \ln P(t,S)}{T-S}.$$

Letting S approach T defines the *instantaneous forward rate* $f(t,T)$,

$$f(t,T) = \lim_{S \to T} f(t,S,T) = -\frac{\partial \ln P(t,T)}{\partial T}. \tag{2.6}$$

In turn the bond price as a function of instantaneous forward rates is given by

$$P(t,T) = \exp\left(-\int_t^T f(t,s)ds\right). \tag{2.7}$$

Equation (2.6) implies that the instantaneous short rate can be written in terms of the instantaneous forward rate as

$$r_t = f(t,t). \tag{2.8}$$

We will deviate from these conventions when we analyze term structure models in discrete time in chapter 3. There, the base unit of time will be one month, and time spans will be integer multiples of that. That is, we have $n = 0, 1, 2 \dots$. Thus, the smallest positive time span considered is $n = 1$. We will write r_t for the one-month yield, i.e. $r_t = y_t^1$, although unlike the definition in (2.5), the time to maturity is not an instant but rather the shortest time interval considered. In the discrete time setting, the so defined r_t will also be referred to as the short rate.

The *term structure* of interest rates at time t is the mapping between time to maturity and the corresponding yield. Thus it can be written as a function ϕ_t with

$$\phi_t : [0, M^*] \to \mathbb{R}, \quad n \to \phi_t(n) = y(t, t+n)$$

where $M^* < \infty$ is an upper bound on time to maturity. The graph of the yield $y(t, t+n)$ against time to maturity n is referred to as the *yield curve.*

Besides continuously compounded rates, it is also common to use simple rates. An important example is the *simply compounded spot rate* $R(t,T)$ which is defined in terms of the zero bond price as

$$R(t,T) = \frac{1}{T-t}\left(\frac{1}{P(T,t)} - 1\right). \tag{2.9}$$

The zero bonds introduced above are the most basic and important ingredient for theoretical term structure modeling. In reality, however, most bonds are *coupon bonds* that make coupon payments at predetermined dates before maturity. We will consider coupon bonds with the following properties. Denote the number of dates for coupon payments after time t by N.[3] They will be indexed by T_1, ..., T_N, where $T_N = T$ coincides with the maturity date of the bond. At T_i, $i = 1, \dots, N-1$, the bond holder receives coupon payments c_i, at T_N he receives coupon payments plus face value, $c_N + 1$.

The payment stream of a coupon bond can be replicated by a portfolio of zero bonds with maturities T_i, $i = T_1, \dots, T_N$. Consequently, the price $P^C(t,T,c)$ of the fixed coupon bond[4] has to be equal to the value of that portfolio. That is, we have

$$P^C(t,T,c) = c_1 P(t,T_1) + \ldots + c_{N-1}P(t,T_{N-1}) + (c_N + 1)P(t,T_N). \tag{2.10}$$

2.2 Data Set and Some Stylized Facts

In this subsection we want to present some of the stylized facts that characterize the term structure of interest rates. Of course those 'facts' may change

[3] Of course, the number of coupon dates until maturity depends on t, but we set $N(t) = N$ for notational simplicity.

[4] $c = \{c_1, \dots, c_N\}$ denotes the sequence of coupon payments.

if using different sample periods or if looking at different countries. However, there are some features in term structure data that are regularly observed for a wide range of subsamples and for different countries.[5] We will base the presentation on an actual data set with US treasury yields. Before we come to the analysis of the data we make a short digression and give an exposition on how data sets of zero coupon yields are usually constructed.

Since yields of zero coupon bonds are not available for each time and each maturity, such data have to be estimated from observed prices of coupon bonds. For the estimation at some given time t, it is usually assumed that the term structure of zero bond prices $P(t, t+n)$ viewed as a function of time to maturity n, can be represented by a smooth function $S(n; \theta)$, where θ is a vector of parameters.[6]

The theoretic relation between the price of a coupon bond and zero bond prices is given by (2.10) above. For the purpose of estimation, each zero bond price on the right hand side of (2.10) is replaced by the respective value of the function $S(n; \theta)$. Thus, e.g., $P(t, T_i) = P(t, t+n_i)$ is replaced by $S(n_i; \theta)$. Now, on the left hand side of the equation there is the observed coupon bond price, whereas the right hand side contains the 'theoretical price' implied by the presumed function $S(n; \theta)$. From a couple of observed coupon bond prices, implying a couple of those equations, the parameters θ can be estimated by minimizing some measure for the overall distance between observed and theoretical prices. Having estimated θ, one can estimate any desired zero bond price at time t as $\hat{P}(t, t+n) = S(n; \hat{\theta})$. Estimated yields are obtained by plugging $\hat{P}$ into (2.2).

As for the function $S(n; \theta)$, it has to be flexible enough to adopt to different shapes of the term structure, but at the same time it has to satisfy some smoothness restrictions. Specific functional forms suggested in the literature include the usage of polynomial splines,[7] exponential splines,[8] and parametric specifications.[9]

The data set used in this book is based on [84] and [20]. It is the same set as used by Duffee [42].[10] The set consists of monthly observations[11] of annual yields for the period of January 1962 to December 1998. The sample contains yields for maturities of 3, 6, 12, 24, 60 and 120 months. Thus, we have 6 time series of 444 observations each. Yields are expressed in percentages, that is

[5] For a more elaborate discussion of statistical properties of term structure data, see [89]. Compare also Backus [11] who analyzes a data set similar to ours.

[6] For a more detailed exposition of the construction of zero bond prices see, e.g., [5] or [26].

[7] See [82] and [83].

[8] See [114].

[9] See [88] and [106].

[10] We obtained it from G. R. Duffee's website http:// faculty.haas.berkeley.edu /duffee/affine.htm.

[11] We write 'observations' but keep in mind that the data are in fact estimated from prices which are truly observable as outlined above.

yields as defined by (2.2) are multiplied by 100. Three of the six time series are graphed in figure 2.1, table 2.1 provides summary statistics of the data.

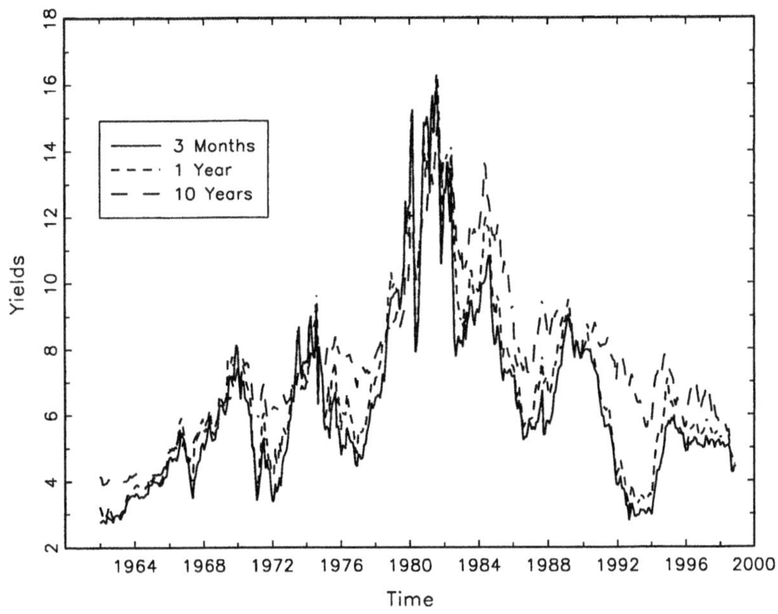

Fig. 2.1. Yields from 01/1962 - 12/1998

Table 2.1. Summary statistics of yields in levels

Mat	Mean	Std Dev	Skew	Kurt	Auto Corr
3	6.32	2.67	1.29	1.80	0.974
6	6.56	2.70	1.23	1.60	0.975
12	6.77	2.68	1.12	1.24	0.976
24	7.02	2.59	1.05	1.02	0.978
60	7.36	2.47	0.95	0.68	0.983
120	7.58	2.40	0.78	0.31	0.987

For each time to maturity (Mat) the columns contain mean, standard deviation, skewness, excess kurtosis, and autocorrelation at lag 1.

As table 2.1 shows, yields at all maturities are highly persistent. The mean increases with time to maturity. Ignoring the three-month yield, the standard deviation falls with maturity. For interpreting the coefficient of skewness and excess kurtosis, note that they should be close to zero if the data are normally distributed.

The means of yields are graphed against the corresponding maturity in figure 2.2. Data are represented by filled circles. The connecting lines are drawn for optical convenience only. The picture shows that the mean yield curve has a concave shape: mean yields rise with maturity, but the increase becomes smaller as one moves along the abscissa.

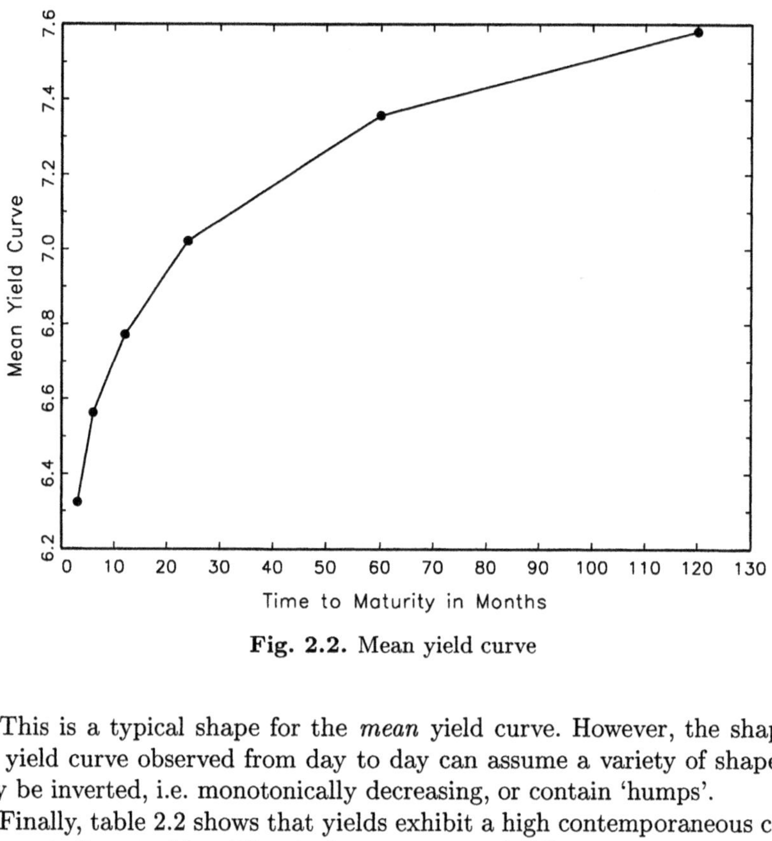

Fig. 2.2. Mean yield curve

This is a typical shape for the *mean* yield curve. However, the shape of the yield curve observed from day to day can assume a variety of shapes. It may be inverted, i.e. monotonically decreasing, or contain 'humps'.

Finally, table 2.2 shows that yields exhibit a high contemporaneous correlation at all maturities. That is, interest rates of different maturities tend to move together.

We now turn from levels to yields in first differences. That is, if $\{y_1^{n_i}, \ldots, y_T^{n_i}\}$ denotes an observed time series of the n_i-month yield in levels, we now consider the corresponding time series $\{\Delta y_2^{n_i}, \ldots, \Delta y_T^{n_i}\}$ with $\Delta y_t^{n_i} = y_t^{n_i} - y_{t-1}^{n_i}$.

Three of the six time series are graphed in figure 2.3. Table 2.3 shows summary statistics of yields in first differences.

Again, the standard deviation falls with time to maturity. The high autocorrelation that we have observed for yields in levels has vanished. Skewness

Table 2.2. Correlation of yields in levels

Mat	3	6	12	24	60	120
3	1.000					
6	0.996	1.000				
12	0.986	0.995	1.000			
24	0.962	0.975	0.990	1.000		
60	0.909	0.924	0.950	0.982	1.000	
120	0.862	0.878	0.908	0.952	0.991	1.000

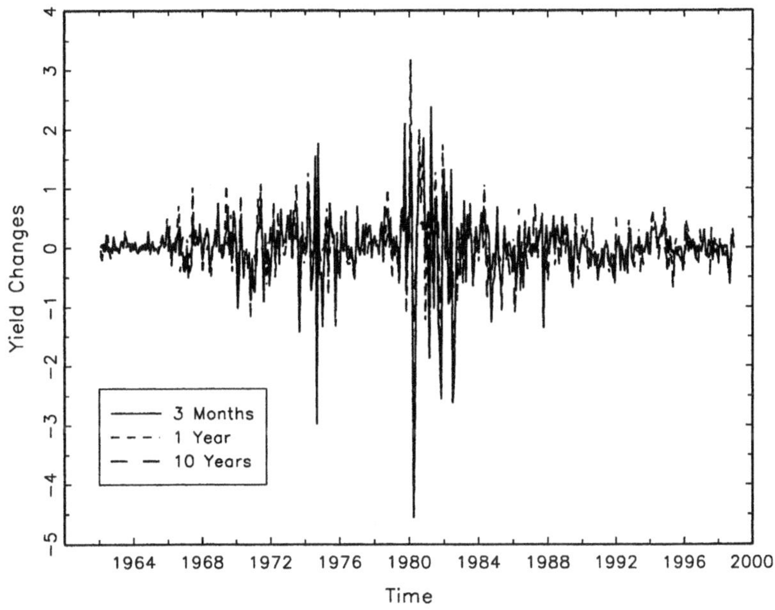

Fig. 2.3. First differences of yields

Table 2.3. Summary statistics of yields in first differences

Mat	Mean	Std Dev	Skew	Kurt	Auto Corr
3	0.0038	0.58	-1.80	14.32	0.115
6	0.0034	0.57	-1.66	15.76	0.155
12	0.0030	0.56	-0.77	12.31	0.158
24	0.0024	0.50	-0.36	10.35	0.146
60	0.0016	0.40	0.12	4.04	0.096
120	0.0015	0.33	-0.11	2.29	0.087

is still moderate but excess kurtosis is vastly exceeding zero. Moreover, excess kurtosis differs with maturity having a general tendency to decrease with it. This leads to the interpretation that especially at the short end of the term structure, extreme observations occur much more often as would be compatible with the assumption of a normal distribution. We will refer back to this remarkable leptokurtosis in chapter 3 where theoretical models are derived, and in chapter 9 which contains an empirical application.

The contemporaneous correlation of differenced yields is also high, as is evident from table 2.4. However, the correlations are consistently lower than for yields in levels.

Table 2.4. Correlation of yields in first differences

Mat	3	6	12	24	60	120
3	1.000					
6	0.952	1.000				
12	0.867	0.957	1.000			
24	0.783	0.887	0.960	1.000		
60	0.645	0.762	0.859	0.936	1.000	
120	0.547	0.659	0.742	0.830	0.934	1.000

For a concrete data set of US interest rates we have presented a number of characteristic features. We will refer to these features in subsequent chapters when we deal with different theoretical models and when we present an empirical application. Many of the features in our data set are part of the properties which are generally referred to as stylized facts characterizing term structure data. However, there are more stylized facts documented in the literature than those reported here. We have said nothing about them since they will not play a role in the following chapters. Moreover, some of the features reported for our data set may vanish when considering different samples or different markets.

3

Discrete-Time Models of the Term Structure

This chapter deals with modeling the term structure of interest rates in a discrete time framework.[1] We introduce the equivalence between the absence of arbitrage opportunities and the existence of a strictly positive pricing kernel. The pricing kernel is also interpreted from the perspective of a multiperiod consumption model. Two one-factor models for the term structure are discussed, one with a normally distributed factor innovation, another with a factor innovation whose distribution is a mixture of normal distributions. These are generalized to the case with multiple factors and the class of affine multifactor Gaussian mixture (AMGM) models is introduced.

3.1 Arbitrage, the Pricing Kernel and the Term Structure

In order to introduce the notion of arbitrage, we use a discrete-time model with N assets.[2] Uncertainty of the discrete time framework is modeled as a probability space $(\Omega, \mathcal{F}, \mathbb{P})$. There are $T^* + 1$ dates indexed by $0, 1, \ldots, T^*$. The sub-sigma algebra $\mathcal{F}_t \subseteq \mathcal{F}$ represents the information available at time t. Accordingly, the filtration $\mathbb{F} = \{\mathcal{F}_0, \mathcal{F}_1, \ldots, \mathcal{F}_{T^*}\}$, with $\mathcal{F}_s \subseteq \mathcal{F}_t$ for $s \leq t$, governs the evolution of information through time. By $\{X_t\} = \{X_0, X_1, \ldots, X_{T^*}\}$ we denote a sequence of random vectors. Such a sequence, also called a random process, is said to be *adapted*, if each component of X_t is a random variable with respect to $(\Omega, \mathcal{F}_t)$. If a random sequence is only defined for a subset of $\{0, 1, \ldots, T^*\}$, this will be explicitly annotated.

---

[1] For surveys of term structure modeling in discrete and continuous time, see the expositions by [99] and [89], the monograph by [5], chapter 4, and the respective sections in the surveys by [105] and [27].

[2] The following exposition leading to theorem 3.1 is based on [64].

The baseline model contains N assets, each of them characterized by an adapted price process $\{P_t^i\}$.[3] Prices at time t are collected in the vector $P_t = (P_{1t}, \dots, P_{Nt})'$. A *trading strategy* $H = \{H_t\}$ is a vector-valued adapted process, where $H_t = (H_{1t}, \dots, H_{Nt})' \in \mathbb{R}^N$, and H_{it} represents the amount of asset i held in a portfolio within the time interval $(t, t+1]$. That is, the portfolio is constructed after obtaining the information at time t and is then held until the end of the following period. Note that H_{it} can be negative, which is interpreted as a short position in asset i at time t. Concerning the borders of the investment horizon, it is further assumed that $H_{-1} = 0$, and $H_{T^*} = 0$.[4]

Associated with the process of portfolio holdings H is a *gain process* $\delta(H) = \{\delta_t(H)\}$, which is defined by

$$\delta_t(H) := H'_{t-1}P_t - H'_tP_t.$$

At time t the investor arrives with portfolio holdings H_{t-1} from the last period. Then prices are revealed, and the investor adjusts his asset holdings. If δ_t is negative, the investor has to bring in capital from outside to finance the new portfolio, if δ_t is positive[5], the new portfolio is 'cheaper' than the old one, and the difference δ_t can be put aside. At $t = 0$, we have $\delta_0 = -H'_0P_0$ which is the withdrawal necessary to finance the initial portfolio. The value of the final portfolio at time $t = T^*$ is given by $\delta_{T^*} = H'_{T^*-1}P_{T^*}$. A trading strategy H with $\delta_t(H) = 0$ for $t = 1, \dots, T^* - 1$ is called self-financing. The process of changes in portfolio positions will be denoted by $\xi = \{\xi_t\}$, i.e. $\xi_t := H_t - H_{t-1}$.

Finally, we introduce the notions of a claim, a hedge, and market completeness.[6] A *claim* $\zeta = \{\zeta_t\}$, $t = 1, \dots, T^*$ is an adapted process. At each time t the owner of such a claim is entitled to the payoff ζ_t. A claim is said to be *hedgeable* (or *replicaple*) if there is a trading strategy H whose gain function satisfies $\delta_t(H) = \zeta_t$ for each t. The model with our N assets is said to be *complete* if every claim can be hedged.

Now we are in a position to define what is meant by arbitrage. The literature contains various definitions, many of them turn out to be equivalent. Basically, the notion of arbitrage refers to a situation in which it is possible to come up with a (dynamic) portfolio that never costs anything but pays off some positive amount with a positive probability. Here we use the definition by [64].

---

[3] We abstract from the possibility of dividend paying assets. The main results are not affected by including dividends. However, in the remainder of the chapter we focus on zero-coupon bonds and may leave the question of dividend payments aside.

[4] Formally, time $t = -1$ is not part of our discrete time scale introduced above. H_{-1} refers to the portfolio holdings an instant before time $t = 0$.

[5] We drop the H argument, when it is clear with which trading strategy δ is associated.

[6] This is not needed for the following considerations concerning arbitrage. However, these concepts will be occasionally referred to below.

A trading strategy is called *arbitrage strategy* (or arbitrage for short) if $\delta_t(H) \geq 0$ for $t = 0, 1, \ldots, T^*$ and $P(\delta_t(H) > 0) > 0$ for at least one t from $\{0, 1, \ldots, T^*\}$.

The following theorem establishes a necessary and sufficient condition for a market to be arbitrage-free. It is this relationship that will serve as the fundament on which all discrete-time term structure models of this chapter will be based.

Theorem 3.1 (Stochastic discount factor and no arbitrage). *The following two statements are equivalent:*

i) There exists no arbitrage strategy.
ii) For each $t = 1, \ldots, T^*$ and each $i = 1, \ldots, N$ there exists an $\mathcal{F}_t$ measurable M_t with $Pr(M_t > 0) = 1$, $E|M_t P_t^i| < \infty$, and

$$P_{t-1}^i = E(M_t P_t^i | \mathcal{F}_{t-1}). \tag{3.1}$$

Proof. See [64], p. 108, et seqq. □

The random variable M_t that plays the central role in the theorem will be referred to as the 'stochastic discount factor' or the 'pricing kernel'. In the following we will sometimes make use of a slightly different representation of the basic asset pricing equation (3.1). Denote by R_t^i the *gross one-period return* of the ith asset, i.e. $R_t^i = P_t^i / P_{t-1}^i$. Thus, in an arbitrage-free market, the return of asset i has to satisfy

$$E_{t-1}(M_t R_t^i) = 1. \tag{3.2}$$

Here and in the following we adopt the short-hand notation for the conditional expectation, $E_t(\cdot) := E(\cdot|\mathcal{F}_t)$.

Note that (3.2) also holds unconditionally. Taking unconditional expectations on both sides of (3.2) yields

$$E(M_t R_t^i) = 1. \tag{3.3}$$

The stochastic discount factor (SDF) M_t may be given an economic interpretation. In the following we will show that (3.2) also results from the first order conditions of an intertemporal utility maximization problem. Consider an agent with preferences over the consumption stream $C = \{C_t\}$ represented by a utility function of the form,

$$U(C) = \sum_{t=0}^{T^*} \beta^t E_0 \left[u(C_t) \right], \tag{3.4}$$

where β is the agent's time discount factor. The individual is entitled to an endowment stream $\{e_t\}$. At each period he can invest in N assets. Resources available at time t consist of the endowment e_t and the portfolio holdings

$H'_{t-1}P_t$. These can be spent on consumption C_t and for constructing a new portfolio that costs H'_tP_t. That is, at each period the agent faces the budget constraint

$$e_t + H'_{t-1}P_t = C_t + H'_tP_t$$

or

$$C_t = e_t - \xi'_tP_t, \tag{3.5}$$

with $\xi_t = H_t - H_{t-1}$ as defined above. The optimal path of consumption and investment, (C^*, ξ^*), is given by the solution of the problem

$$\begin{aligned} &\max_{C,\xi} U(C), \\ \text{s.t.} \quad &C_t = e_t - \xi'_tP_t, \\ &C_t \geq 0, \quad t = 0, 1, \ldots, T^*. \end{aligned}$$

Assuming an interior solution, one obtains from the first order conditions

$$E_t\left(\beta\frac{u'(C_{t+1})}{u'(C_t)} \cdot \frac{P^i_{t+1}}{P^i_t}\right) = 1, \quad i = 1, \ldots, N. \tag{3.6}$$

But this is just our pricing relation (3.2) from above:[7] the stochastic discount factor is given by the ratio of marginal utilities,

$$M_{t+1} = \frac{\beta u'(C_{t+1})}{u'(C_t)}. \tag{3.7}$$

Thus, M_{t+1} can be characterized as the intertemporal rate of substitution between consumption at time t and time $t+1$. Assuming monotonicity of the utility function, the so defined pricing kernel is strictly positive, a condition that theorem 3.1 requires for the absence of arbitrage.[8]

Before we turn to exploiting the relationship (3.1) for building models of the term structure of interest rates, some comments on the stochastic discount factor in general are in order.

To begin with, theorem 3.1 states that for the no arbitrage condition to hold, it is necessary that a stochastic discount factor exists. However, it is not required that it has to be unique. It turns out that the pricing kernel, given one exists, is unique if and only if the model is complete.

Consider a one-period riskfree asset whose price at time t is P^f_t and whose price at time $t+1$ is $P^f_{t+1} = 1$. The associated one-period rate of return is given by $R^f_{t+1} := 1/P^f_t$. From (3.2) this is linked to the conditional mean of the discount factor as

[7] Here written for period $t+1$ instead of t.

[8] See [43] for a discussion of the relation between individual agent optimality, equilibrium with multiple agents, Pareto optimality and no arbitrage.

$$E_t(M_{t+1}) = \frac{1}{R^f_{t+1}}. \tag{3.8}$$

Denote by $Z^i_t := R^i_t - R^f_t$ the excess return of asset i over the riskfree rate. Then straightforward manipulation of (3.2) leads to the relationship

$$E_t(Z^i_{t+1}) = \frac{-Cov_t(M_{t+1}, Z^i_{t+1})}{E_t(M_{t+1})}. \tag{3.9}$$

Thus, expected excess returns are determined by its covariance with the discount factor. An asset return whose covariance with the SDF is negative and large in absolute value has a high excess return. Turning back to the economic interpretation (3.6), a large positive value of the SDF implies high marginal utility, that is a state of low consumption. Due to its negative correlation with the SDF, the described asset generates low returns in a situation in which each additional unit of payoff would be extremely valuable to consumers. Thus, the high average excess return can be interpreted as a compensation for such a 'cyclical' behavior of the asset.[9]

Finally, according to Cochrane [33], p. xv, the majority of asset pricing models can be formulated within the stochastic discount factor framework. He summarizes the prototypical specification of any model as a set of two equations. One is the basic pricing equation (3.1), the other specifies the SDF as a function of some explanatory variables and parameters. In light of the consumption based explanation, any specification of the discount factor can be interpreted as a proxy for marginal utility. For instance, the famous capital asset pricing model results from the specification[10]

$$M_t = a - bR^m_t,$$

where R^m_t is the return of the market or wealth portfolio.

We will use the pricing kernel approach for specifying models of the term structure of interest rates. Consider a zero bond at time t that has n periods left until maturity, and denote its price by P^n_t.[11] In the next period, this bond has only $n-1$ periods left until maturity and a price of P^{n-1}_{t+1}. The two prices are related by the no-arbitrage condition (3.1). We have

$$P^{n+1}_t = E_t(M_{t+1}P^n_{t+1}). \tag{3.10}$$

The price of a bond with one period to maturity ($n = 1$) can be written directly in terms of next period's SDF. First recall that $P^0_{t+1} = 1$ for all t since the zero bond pays off one unit at maturity. Therefore

$$P^1_t = E_t(M_{t+1}). \tag{3.11}$$

---

9 See [27].

10 See [33].

11 In the discrete time context we can think of one period being one month.

As already seen in (3.8), the gross return of the one-period bond equals $1/E_t(M_{t+1})$. The one-period yield is given by

$$r_t := y_t^1 = -\ln E_t(M_{t+1}).$$

By means of repeated substitution we can also write the prices of longer term bonds, i.e. those with $n > 1$, in terms of future discount factors. For the two-period bond:

$$\begin{aligned} P_t^2 &= E_t(M_{t+1}P_{t+1}^1) \\ &= E_t(M_{t+1}E_{t+1}(M_{t+2})) \\ &= E_t(E_{t+1}(M_{t+1}M_{t+2})) \\ &= E_t(M_{t+1}M_{t+2}), \end{aligned}$$

where the second equality follows from (3.11) and the fourth follows from the law of iterated expectations. Proceeding in the same fashion, one obtains for general n:

$$P_t^n = E_t(M_{t+1}M_{t+2}\dots M_{t+n}). \tag{3.12}$$

Equation (3.12) can be viewed as a recipe for constructing a term structure model. One has to specify a stochastic process for the SDF with the property that $P(M_t > 0) = 1$ for all t. The term structure of zero bond prices, or equivalently yields, at time t is then obtained by taking conditional expectations of the respective products of subsequent SDFs, given the information at time t. The models we consider in the following, however, will use a different construction principle, that we now summarize.[12]

The construction of a term structure model starts with a specification of the pricing kernel. The logarithm of the SDF is linked to a vector of state variables X_t via a function $f(\cdot)$ and a vector of innovations u_{t+1},

$$\ln M_{t+1} = f(X_t) + u_{t+1},$$

such that the conditional mean of $\ln M_{t+1}$ equals $f(X_t)$. The specification of the evolution of X_t and u_t determines the evolution of the SDF. Positivity of the SDF is guaranteed by modeling its logarithm rather than its level. Then a solution for bond prices is proposed, i.e. we guess a functional relationship

$$P_t^n = f_P(X_t, C_n, t), \tag{3.13}$$

where the vector C_n contains maturity dependent parameters. This trial function is plugged in for P_t^{n+1} and P_{t+1}^n, respectively, on the left hand side and the right hand side of (3.10). If parameters C_n can be chosen in such a way

[12] See [33], [26] or [27]. The following exposition summarizes the basic idea in a somewhat heuristic fashion. In the next sections the steps will be put into concrete terms when particular term structure models are constructed using this building principle.

that (3.10) holds as an identity for all t and n, then our guess for the solution function has been correct.[13] Typically, the parameters C_n will depend on the parameters governing the relation between the state vector and the SDF as well as on those parameters showing up in the law of motion of the state vector and the innovation.

Summing up, we need three ingredients for a term structure model: a specification for the evolution of the state vector X_t and the innovations u_t, a relationship between X_t, u_t and the log SDF, and finally the fundamental pricing relation (3.10).

The models considered in this book will have the convenient property that the solution function f_P for bond prices in (3.13) is exponentially affine in the state vector. That is, we will have

$$P_t^n = e^{-A_n - B_n' X_t},$$

where A_n and B_n are coefficient functions that depend on the model parameters. Accordingly, using (2.2), yields are affine in X_t,

$$y_t^n = \frac{1}{n}\left(A_n + B_n' X_t\right). \tag{3.14}$$

As evident from (3.14), the state vector X_t drives the term structure of zero coupon yields: the dynamic properties of X_t induce the dynamic properties of y_t^n. However, we have said nothing specific about the nature of X_t yet. By the comments made above, all variables that enter the specification of the SDF can be interpreted as proxies for marginal utility growth. In specific economic models, the components of X_t may have some concrete economic interpretation as, e.g., the market portfolio in the CAPM. For the following, however, we will refer to X_t simply as a vector of 'factors' without attaching a deeper interpretation to them. Accordingly, in the term structure models considered in this book, the factors will be treated as latent variables. This implies that the focus of this book is on "relative" pricing as Cochrane[14] calls it: bond yields in our models will satisfy the internal consistency condition of absence of arbitrage opportunities, but we will not explore the deeper sources of macroeconomic risk as the ultimate driving forces of bond yield dynamics.[15]

The last comment in this subsection discusses the distinction between nominal and real stochastic discount factors. Up to now we have talked about payoffs and prices but we have said nothing about whether they are in real

---

[13] This is the same principle as with the method of undetermined coefficients which is popular for solving difference equations or differential equations. In fact, (3.10) is a difference equation involving a conditional expectation.

[14] See [33].

[15] Quite recent literature is linking arbitrage-free term structure dynamics to small macroeconomic models, see, e.g., [6], [62] and [95]. The link between macroeconomic variables and the short-term interest rate is typically established by a monetary policy reaction function.

terms or in nominal terms. That is, does a zero bond pay off one dollar at maturity or one unit of consumption? Going back to the derivation of the basic pricing formula (3.10) from a utility maximizing agent, we had the price of the consumption good being normalized to one. That is, asset prices and payoffs were all in terms of the consumption good. Specifically, all bonds in this set up would be real bonds. When we aim to apply our pricing framework to empirical data, however, most of the traded bonds have their payoffs specified in nominal terms.[16] Thus, the natural question arises whether formula (3.10) can also be used to price nominal bonds. It turns out that the answer is yes, and in that case the SDF has to be interpreted as a nominal discount factor.[17]

To see this, we start with equation (3.10) with the interpretation that prices are expressed in units of the consumption good. Let q_t^* denote the time t price of one unit of the good in dollars.[18] Then our real prices in (3.10) are connected to nominal prices, marked by a \$ superscript, as

$$P_t^{\$n} = q_t^* \cdot P_t^n.$$

Solving for P_t^n and plugging this into (3.10) yields

$$\frac{P_t^{\$n}}{q_t^*} = E_t\left(M_{t+1}\frac{P_{t+1}^{\$\,n-1}}{q_{t+1}^*}\right), \tag{3.15}$$

equivalently,

$$P_t^{\$n} = E_t\left(M_{t+1}\frac{q_t^*}{q_{t+1}^*}P_{t+1}^{\$\,n-1}\right). \tag{3.16}$$

The ratio q_{t+1}^*/q_t^* is the gross rate of inflation from period t to $t+1$, which will be denoted by $1+\Pi_{t+1}$. Thus, if we define the nominal stochastic discount factor $M_{t+1}^{\$}$ by

$$M_{t+1}^{\$} = \frac{M_{t+1}}{1+\Pi_{t+1}} \tag{3.17}$$

we can write

$$P_t^{\$n} = E_t\left(M_{t+1}^{\$}P_{t+1}^{\$\,n-1}\right). \tag{3.18}$$

which has the same form as (3.10). There, real prices were related by a real pricing kernel, whereas here, nominal prices are related by a nominal pricing kernel. For the following we will not explicitly denote which version of the basic pricing equation (nominal or real) we refer to. That is, we may use equation (3.18) for nominal bonds, but drop the \$ superscript. Finally, we remark that in face of (3.17), every model for the evolution of the nominal SDF imposes a joint restriction of the dynamic behavior of the real SDF and inflation.

---

[16] Of course, the problem is alleviated a little when working with index-linked bonds, the nominal payoffs of which depend on some index of inflation. Those bonds may proxy for real bonds in empirical applications.

[17] See [26] and [33] for the following exposition.

[18] We may also talk about a bundle of consumption goods and q_t^* being a (consumer) price index.

3.2 One-Factor Models

In this section we consider discrete-time term structure models for which the state variable is a scalar. Although these one-factor models turn out to be unsatisfactory when confronted with empirical data, it is nevertheless worthwhile to explore the properties of these models in some detail, since they can be interpreted as cornerstones from which more elaborate extensions such as multifactor models are developed.

3.2.1 The One-Factor Vasiček Model

The first model under consideration can be viewed as a discrete-time analogy to the famous continuous-time model of Vasiček [113]. Therefore, we also refer to the discrete time version as the (discrete-time) Vasiček model.[19]

According to our general specification scheme outlined above, we start with a specification of the pricing kernel. The negative logarithm of the SDF is decomposed into its conditional expectation $\delta + X_t$ and a zero mean innovation w_{t+1},

$$-\ln M_{t+1} = \delta + X_t + w_{t+1}. \tag{3.19}$$

The state variable X_t is assumed to follow a stationary AR(1) process with mean θ, autorogressive parameter κ and innovation u_t,

$$X_t = \theta + \kappa(X_{t-1} - \theta) + u_t. \tag{3.20}$$

Innovations to the SDF and the state variable may be contemporaneously correlated. This correlation can be parameterized as

$$w_{t+1} = \lambda u_{t+1} + v_{t+1},$$

where v_{t+1} is uncorrelated with u_{t+1}. Accordingly, the covariance between w_t and u_t is proportional to λ. Replacing w_{t+1} in (3.19) by this expression yields

$$-\ln M_{t+1} = \delta + X_t + \lambda u_{t+1} + v_{t+1}.$$

It turns out that deleting v_{t+1} from the latter equation leads to a parallel down shift of the resulting yield curve. However, such a level effect can be compensated for by increasing the parameter δ appropriately. Neither the dynamics nor the shape of the yield curve are affected.[20] Therefore, v_t will be dropped from the model leaving the modified equation

$$-\ln M_{t+1} = \delta + X_t + \lambda u_{t+1}. \tag{3.21}$$

---

[19] Our description of the one-factor models is based on [11].

[20] One could easily show this assertion by conducting the subsequent derivation of the term structure equation with v_{t+1} remaining in the model.

The distribution of the state innovation is assumed to be a Gaussian white noise sequence:

$$u_t \sim i.i.d.\, N(0, \sigma^2). \tag{3.22}$$

The model equations (3.20), (3.21), (3.22), are completed by the basic pricing relationship (3.10). In the following we will make use of its logarithmic transformation

$$-\ln P_t^{n+1} = -\ln E_t(M_{t+1} P_{t+1}^n). \tag{3.23}$$

We are now ready to solve the model. That is, zero bond prices, or equivalently yields, will be expressed as a function of the state variable X_t. According to the next result yields of zero bonds are affine functions of the state variable X_t.

Proposition 3.2 (Yields in the discrete-time Vasiček model). *For the one-factor Vasiček model (3.20) - (3.22) zero bond yields are given as*

$$y_t^n = \frac{A_n}{n} + \frac{B_n}{n} X_t \tag{3.24}$$

with[21]

$$B_n = \sum_{i=0}^{n-1} \kappa^i = \frac{1-\kappa^n}{1-\kappa}, \tag{3.25}$$

$$A_n = \sum_{i=0}^{n-1} G(B_i), \tag{3.26}$$

where

$$G(B_i) = \delta + B_i\, \theta(1-\kappa) - \frac{1}{2}(\lambda + B_i)^2 \sigma^2.$$

Proof. The structure of the proof follows the derivation from [11]. One starts by guessing that bond prices satisfy

$$\ln P_t^n = -A_n - B_n X_t \tag{3.27}$$

and then shows that in fact the no arbitrage condition (3.23) is satisfied if B_n and A_n are given by (3.25) and (3.26), respectively.

The right hand side of (3.23) can be written as $-\ln E_t \left(e^{\ln M_{t+1} + \ln P_{t+1}^n}\right)$. Using (3.21) and the guess (3.27), the exponent is given by

$$\begin{aligned}
&\ln M_{t+1} + \ln P_{t+1}^n \\
&= -\delta - X_t - \lambda u_{t+1} - A_n - B_n X_{t+1} \\
&= -\delta - X_t - \lambda u_{t+1} - A_n - B_n(\theta(1-\kappa) + \kappa X_t + u_{t+1}) \\
&= -\delta - A_n - X_t - B_n(\theta(1-\kappa) + \kappa X_t) - (\lambda + B_n) u_{t+1} \\
&=: V_{t+1}
\end{aligned}$$

[21] Empty sums are evaluated as zero.

Conditional on $\mathcal{F}_t$, the information given at time t, this expression is normally distributed with mean

$$E_t(V_{t+1}) = -\delta - A_n - B_n\,\theta(1-\kappa) - (1+\kappa B_n)X_t,$$

and variance

$$Var_t(V_{t+1}) = (\lambda + B_n)^2\sigma^2.$$

Thus, the conditional distribution of $e^{\ln M_{t+1} + \ln P^n_{t+1}}$ is log-normal with mean

$$\begin{aligned} &E_t(e^{\ln M_{t+1} + \ln P^n_{t+1}}) \\ &= \exp\left[(-\delta - A_n - B_n\,\theta(1-\kappa) - (1+\kappa B_n)X_t) + \frac{1}{2}\left((\lambda + B_n)^2\sigma^2\right)\right]. \end{aligned}$$

Taking the negative logarithm of this expression, one obtains the right hand side of (3.23). Using the guess (3.27) again for the left hand side of (3.23) one ends up with the equation

$$A_{n+1} + B_{n+1}X_t = \delta + A_n + B_n\,\theta(1-\kappa) + (1+\kappa B_n)X_t - \frac{1}{2}(\lambda + B_n)^2\sigma^2.$$

Collecting terms, the equation can be formulated as $\tilde{c}_1 + \tilde{c}_2 X_t = 0$ with

$$\begin{aligned} \tilde{c}_1 &= A_{n+1} - A_n - \delta - B_n\,\theta(1-\kappa) + \frac{1}{2}(\lambda + B_n)^2\sigma^2, \\ \tilde{c}_2 &= B_{n+1} - 1 - \kappa B_n. \end{aligned}$$

To guarantee the absence of arbitrage opportunities, this relation has to be satisfied for all values of X_t. Thus, we must have $\tilde{c}_1 = \tilde{c}_2 = 0$, leading to the conditions

$$A_{n+1} = A_n + \delta + B_n\,\theta(1-\kappa) - \tfrac{1}{2}(\lambda + B_n)^2\sigma^2, \tag{3.28}$$

$$B_{n+1} = 1 + \kappa B_n. \tag{3.29}$$

This is a system of difference equations in A_n and B_n. Using $P^0_t = 1$ in (3.27) one obtains the additional constraint

$$-\ln P^0_t = 0 = A_0 + B_0 X_t,$$

that leads to the initial condition

$$A_0 = 0, \quad B_0 = 0 \tag{3.30}$$

for our system of difference equations. Solving (3.28) - (3.30), for A_n and B_n one obtains the expressions given in (3.25) and (3.26). □

Concerning the parameterization of (3.21), we follow [11] and set δ in such a way that X_t coincides with the one-period short rate. According to (3.24) - (3.26) we have

$$y_t^1 = A_1 + B_1 X_t = \delta - \frac{1}{2}\lambda^2\sigma^2 + X_t.$$

Setting

$$\delta = \frac{1}{2}\lambda^2\sigma^2 \tag{3.31}$$

equates X_t with y_t^1, leaving the model to be parameterized in the four parameters $(\theta,\ \kappa,\ \sigma,\ \lambda)$ only.[22]

The simple model that we have introduced is a full description of the term structure of interest rates with respect to both its cross-sectional properties and its dynamics. For any given set of maturities, say $\{n_1, \ldots, n_k\}$, the evolution of the corresponding vector of yields, $(y_t^{n_1}, \ldots, y_t^{n_k})'$ is fully specified by the specification of the factor evolution, (3.20) and (3.22), and the relation between the factor and yields (3.24).

Next, we derive the Sharpe ratio of zero bonds implied by the model. Define the gross one-period return of the n-period bond as

$$R_{t+1}^n = \frac{P_{t+1}^{n-1}}{P_t^n}.$$

For $n = 1$ this is the riskless return to which the special notation R_{t+1}^f has been assigned above. Small characters denote logarithms, i.e.

$$\ln R_{t+1}^n =: r_{t+1}^n, \quad \ln R_{t+1}^f =: r_{t+1}^f.$$

Note that r_t^f is equal to the short rate y_t^1. The Vasiček model implies that the expected one-period excess log-return of the n-period bond over the short rate is[23]

$$E_t(r_{t+1}^n - r_{t+1}^f) = -B_{n-1}\lambda\sigma^2 - \frac{1}{2}B_{n-1}^2\sigma^2. \tag{3.32}$$

For the conditional variance of the excess return we have

$$Var_t(r_{t+1}^n - r_{t+1}^f) = B_{n-1}^2\sigma^2,$$

thus the conditional Sharpe ratio of the n-period bond, defined as the ratio of expected excess return and its standard deviation, turns out to be

$$SR_t^n = \frac{-B_{n-1}\lambda\sigma^2 - \frac{1}{2}B_{n-1}^2\sigma^2}{B_{n-1}\sigma} = -\lambda\sigma - \frac{1}{2}B_{n-1}. \tag{3.33}$$

In the following we will refer to λ (strictly speaking $-\lambda$) as the market price of risk parameter. For any n, given the other model parameters, an increase in

[22] Below, we will introduce a class of multifactor models of which the Vasiček model is a special case. There, a canonical parameterization will be introduced. This, however is not relevant for the subsequent analysis in this section, so we stick to the parameterization that has been introduced just now.

[23] See section C.1 in the appendix.

$-\lambda$ increases the excess return of the bonds. Alternatively, $-\lambda$ is the additional excess return that the bond delivers if its volatility is increased by one unit.

The mean yield curve implied by the model can be computed as

$$\begin{aligned}
E(y_t^n) &= \frac{A_n}{n} + \frac{B_n}{n}\theta \\
&= \frac{1}{n}\left(n\delta + \theta(n - \frac{1-\kappa^n}{1-\kappa}) - \frac{1}{2}\sigma^2 \sum_{i=0}^{n-1}(\lambda + B_i)^2\right) + \frac{1}{n}\frac{1-\kappa^n}{1-\kappa}\theta \\
&= \delta + \theta - \frac{1}{2}\sigma^2 \frac{1}{n}\sum_{i=0}^{n-1}(\lambda + B_i)^2. \qquad (3.34)
\end{aligned}$$

The stylized facts suggest that on average the term structure is upward sloping. It can be shown that for yields to be monotonic increasing in n, the parameter λ has to be sufficiently negative.

Further properties of the discrete-time Vasiček model are discussed in [26]. They derive the characteristics of the implied forward-rate curve and point out that the model is flexible in the sense that it can give rise to an upward sloping, a downward sloping ('inverted') or hump-shaped forward-rate curve. Moreover the consumption based interpretation of the one-factor model is discussed. It turns out that the model implies that expected consumption growth follows an AR(1) whereas realized consumption growth follows an ARMA(1,1).[24]

3.2.2 The Gaussian Mixture Distribution

In the subsequent sections, we will introduce term structure models whose innovation is not Gaussian but is distributed as a mixture of normal distributions. This is why we make a little digression at this point and introduce this distribution and some of its properties which will be needed in the remainder of this book.[25]

A random variable X is said to have a Gaussian mixture distribution with B components if its density can be written as

$$\begin{aligned}
p(x) &= \sum_{b=1}^{B} \omega_b\, \phi(x;\, \mu_b, \sigma_b^2) \qquad (3.35)\\
&= \sum_{b=1}^{B} \omega_b \frac{1}{\sqrt{2\pi\sigma_b^2}} \exp\left(-\frac{1}{2}\frac{(x-\mu_b)^2}{\sigma_b^2}\right), \qquad (3.36)
\end{aligned}$$

$$\text{with} \quad 0 \le \omega_b \le 1, \quad b = 1, \ldots, B, \quad \sum_{b=1}^{B} \omega_b = 1. \qquad (3.37)$$

[24] This is shown for the case with a power utility function. That is, the period utility function $u(C_t)$ in (3.4) has the functional form $u(C_t) = (C_t^{1-\gamma} - 1)/(1-\gamma)$.

[25] For an extensive treatment of finite mixture models see [85] or [111].

Here and in the following $\phi(x;\mu,\sigma^2)$ denotes the density function of the normal distribution $N(\mu,\sigma^2)$ evaluated at x.

The density (3.35) is a weighted sum of B normal density functions each characterized by its own pair (μ_b,σ_b^2) of mean and variance parameters. The ω_b will be called the weights, and the densities $\phi(x;\ \mu_b,\sigma_b^2)$ will be referred to as the component densities. Obviously, the mixture density coincides with a simple normal density if either all pairs (μ_b,σ_b^2) are equal, or if one of the weights equals 1 and thus the others are zero.

The mixture distribution can be interpreted as generating a realization x of X as the outcome of a two-stage process. First, one of B populations or regimes is drawn, where the probability of drawing the bth population is equal to ω_b. Then a draw from the simple normal distribution belonging to that population, i.e. the one characterized by (μ_b,σ_b^2) generates the outcome x.

One can derive the distribution function of X by rewriting $P(X\le x)$ making use of the two-stage interpretation. Let I denote the indicator variable that can take on numbers $1,\ldots,B$ indicating which regime prevails, thus $P(I=b)=\omega_b$. We have

$$\begin{aligned} F(x) &= P(X\le x)\\ &= \sum_{b=1}^{B} P(X\le x, I=b)\\ &= \sum_{b=1}^{B} P(I=b)\,P(X\le x|I=b)\\ &= \sum_{b=1}^{B} \omega_b \Phi\left(\frac{x-\mu_b}{\sigma_b}\right), \end{aligned}$$

where here and in the following $\Phi(x)$ denotes the cumulative distribution function of the standard normal distribution evaluated at x.

We introduce the notation

$$X\sim\sum_{b=1}^{B}\omega_b N(\mu_b,\sigma_b^2),$$

to denote that X has the Gaussian mixture distribution with density as shown above. It is important to note that this does not mean that X is a sum of normally distributed random variables. If this was the case, X would itself be normally distributed.

The Gaussian mixture distribution has the flexibility to take on a wide variety of shapes while keeping a rather parsimonious parameterization.[26]

[26] A collection of differently shaped densities together with the corresponding parameterizations is given in [85].

Already with $B = 2$ components the density can be skewed, bimodal or both. Due to the flexibility of (Gaussian) mixture distributions, they are frequently used in statistics as an approximation of other densities. In a Bayesian context, for instance, the prior density may be written as a mixture of conjugates.[27]

For the analysis of the term structure models, the first four central moments of the mixture distribution will be needed.[28] By straightforward computation it follows that the mean is given as a weighted sum of the component means,

$$E(X) = \sum_{b=1}^{B} \omega_b \mu_b =: \mu. \tag{3.38}$$

The variance is the weighted sum of the component variances plus a term that captures the deviations of the component means from the overall mean:

$$\begin{aligned} &Var(X) \\ &= \int (x-\mu)^2 \left[\sum_{b=1}^{B} \omega_b \phi(x;\, \mu_b, \sigma_b^2)\right] dx \\ &= \sum_{b=1}^{B} \omega_b \left[\int (x - \mu_b + \mu_b - \mu)^2 \phi(x;\, \mu_b, \sigma_b^2) dx\right] \\ &= \sum_{b=1}^{B} \omega_b \left[\int \left((x-\mu_b)^2 + (\mu_b - \mu)^2 + 2(x-\mu_b)(\mu_b-\mu)\right) \phi(x;\, \mu_b, \sigma_b^2) dx\right] \\ &= \sum_{b=1}^{B} \omega_b [\sigma_b^2 + (\mu_b - \mu)^2] =: \sigma^2 \end{aligned} \tag{3.39}$$

The third and fourth moments can be computed in a similar fashion. They are given by

$$E[(X-\mu)^3] = \sum_{b=1}^{B} \omega_b [3(\mu_b - \mu)\sigma_b^2 + (\mu_b - \mu)^3] \tag{3.40}$$

and

$$E[(X-\mu)^4] = \sum_{b-1}^{B} \omega_b [3\sigma_b^4 + 6(\mu_b - \mu)^2 \sigma_b^2 + (\mu_b - \mu)^4], \tag{3.41}$$

respectively.

In the models considered in the remainder of the book, we will often deal with mixtures the components of which have different variances but the same mean. For this case, i.e. if $\mu_b = \mu$ for all b, the formulas for the moments simplify as follows:

[27] See [115].

[28] See, e.g., [67] for the following formulas.

$$Var(X) = \sum_{b=1}^{B} \omega_b \sigma_b^2$$
$$E[(X-\mu)^3] = 0$$
$$E[(X-\mu)^4] = 3\sum_{b=1} \omega_b \sigma_b^4$$

For the Gaussian mixture distribution, the coefficients of skewness and kurtosis, defined by

$$skew(X) = \frac{E[(X-\mu)^3]}{\sigma^3} \quad \text{and} \quad kurt(X) = \frac{E[(X-\mu)^4]}{\sigma^4} - 3,$$

respectively, can be computed using the moments given above. Note that for the simple normal distribution both of them are equal to zero.

We also point out that for the mixture with $\mu_b = \mu$, the coefficient of kurtosis is positive. That is, we have excess kurtosis in comparison to the simple normal distribution. This can be seen as follows. By Jensen's inequality[29]

$$\sum_{b=1}^{B} \omega_b \sigma_b^4 \geq \left(\sum_{b=1}^{B} \omega_b \sigma_b^2 \right)^2 .$$

Rearranging yields

$$\frac{\sum_{b=1}^{B} \omega_b \sigma_b^4}{(\sum_{b=1}^{B} \omega_b \sigma_b^2)^2} \geq 1.$$

This is equivalent to

$$\frac{3\sum_{b=1}^{B} \omega_b \sigma_b^4}{(\sum_{b=1}^{B} \omega_b \sigma_b^2)^2} - 3 \geq 0,$$

where the left hand side is the coefficient of kurtosis which proves the assertion.

Let now X denote a random $g \times 1$ vector. The multivariate Gaussian mixture density has the form

$$p(x) = \sum_{b=1}^{B} \omega_b \, \phi(x;\, \mu_b, Q_b)$$
$$= \sum_{b=1}^{B} \omega_b \frac{1}{\sqrt{(2\pi)^g |Q_b|}} \exp\left(-\frac{1}{2}(x-\mu_b)' Q_b^{-1}(x-\mu_b) \right), \quad (3.42)$$

$$\text{with} \quad 0 \leq \omega_b \leq 1, \quad b = 1, \ldots, B, \quad \sum_{b=1}^{B} \omega_b = 1.$$

[29] Jensen's inequality states that for a convex function $f(w)$, $f(\sum a_b x_b) \leq \sum a_b f(x_b)$ where $a_b \geq 0$, $\sum a_b = 1$. This result applies here with $f(w) = w^2$, $a_b = \omega_b$, and $x_b = \sigma^2$.

The μ_b and Q_b are vectors and matrices of dimension $g \times 1$ and $g \times g$ respectively. Similarly to the univariate case we write

$$X \sim \sum_{b=1}^{B} \omega_b N(\mu_b, Q_b),$$

if X has the described density.

The mean vector and variance-covariance matrix of X are given by

$$E(X) = \sum_{b=1}^{B} \omega_b \mu_b \tag{3.43}$$

and

$$Var(X) = \sum_{b=1}^{B} \omega_b \left[Q_b + (\mu_b - \mu)(\mu_b - \mu)'\right] \tag{3.44}$$

respectively

The marginal densities derived from the multivariate normal mixture are themselves mixtures. Let $x' = (z', y')$ have a multivariate Gaussian mixture distribution,

$$p(x) = p\begin{pmatrix} z \\ y \end{pmatrix} = \sum_{b=1}^{B} \omega_b \phi\left(\begin{pmatrix} z \\ y \end{pmatrix}; \begin{pmatrix} \mu_b^z \\ \mu_b^y \end{pmatrix}, \begin{pmatrix} Q_b^z & Q_b^{zy} \\ Q_b^{yz} & Q_b^y \end{pmatrix}\right) \tag{3.45}$$

then

$$\begin{aligned} p(z) &= \int p(x) dy \\ &= \sum_{b=1}^{B} \omega_b \int \phi\left(\begin{pmatrix} z \\ y \end{pmatrix}; \begin{pmatrix} \mu_b^z \\ \mu_b^y \end{pmatrix}, \begin{pmatrix} Q_b^z & Q_b^{zy} \\ Q_b^{yz} & Q_b^y \end{pmatrix}\right) dy \\ &= \sum_{b=1}^{B} \omega_b \phi(z;\, \mu_b^z, Q_b^z), \end{aligned}$$

where the last equality follows from a standard property of the multivariate normal distribution.

An affine transformation of a random vector that has a Gaussian mixture distribution is also distributed as a mixture of normals. Let X be distributed as a normal mixture,

$$X \sim \sum_{b=1}^{B} \omega_b N(\mu_b, Q_b).$$

Define the random vector Y by

$$Y = LX + l,$$

where L and l are an invertible $g \times g$ matrix and a $g \times 1$ vector of constants, respectively. Then Y is distributed as

$$Y \sim \sum_{b=1}^{B} \omega_b N(L\mu_b + l,\ LQ_bL').$$

This can be seen as follows. We have $X = L^{-1}(Y - l)$. The Jacobian of the transformation is given by L^{-1}. Then by a standard result concerning the distribution of transformed continuous random variables,[30] the density of Y is given by

$$\begin{aligned} & p_Y(y) \\ &= \left|L^{-1}\right| \cdot p_X(L^{-1}(y-l)) \\ &= \sum_{b=1}^{B} \omega_b \cdot \left|L^{-1}\right| \frac{1}{(2\pi)^{g/2}\sqrt{|Q_b|}} e^{-\frac{1}{2}\left(L^{-1}(y-l)-\mu_b\right)' Q_b^{-1}\left(L^{-1}(y-l)-\mu_b\right)}. \end{aligned}$$

The term appearing on the right hand side (everything to the right of the ω_b) is the density of a random variable $V = LW + l$, where W is the simple normal $W \sim N(\mu_b, Q_b)$. Thus,

$$p_Y(y) = \sum_{b=1}^{b} \omega_b\, \phi(y;\ L\mu_b, LQ_bL'),$$

which proves the assertion.

If the vector X is distributed as a simple multivariate normal, diagonality of the variance-covariance matrix implies independence among the entries of X. In case of a multivariate mixture, however, diagonal variance-covariance matrices of the component densities do not imply independence. To see this, consider the bivariate case, i.e. z and y in $x' = (z', y')$ are scalars. With $Q_b^{zy} = Q_b^{yz} = 0$ for all b in (3.45), the joint density is

$$p(x) = \sum_{b=1}^{B} \omega_b \cdot \phi(z;\ \mu_b^z, Q_b^z) \cdot \phi(y;\ \mu_b^y, Q_b^y),$$

which is clearly different from the product of the marginal densities

$$p(z) \cdot p(y) = \left(\sum_{b=1}^{B} \omega_b \phi(z;\ \mu_b^z, Q_b^z)\right) \cdot \left(\sum_{b=1}^{B} \omega_b \phi(y;\ \mu_b^y, Q_b^y)\right).$$

This result is intuitive if one thinks in the two-stage setting described above. *Given* a regime b, z and y are independent. The draw of the regime itself, however, always affects both random variables.

[30] See, e.g., [86].

The last property refers to an exponential transformation. Let X be a g-dimensional random vector with the multivariate mixture of normals distribution, $X \sim \sum_{b=1}^{B} \omega_b N(\mu_b, Q_B)$, and let $c \neq 0$ be a g-dimensional vector of real constants. Then

$$E(e^{c'X}) = \sum_{b=1}^{B} \omega_b \, e^{c'\mu_b + \frac{1}{2}c'Q_b c}. \tag{3.46}$$

The proof makes use of the result from the simple normal. We have

$$\begin{aligned} &E(e^{c'X}) \\ &= \int e^{c'x} \left(\sum_{b=1}^{B} \omega_b \phi(x;\, \mu_b, Q_b) \right) dx \\ &= \sum_{b=1}^{B} \int e^{c'x} \, \phi(x;\, \mu_b, Q_b) dx. \end{aligned}$$

The bth integral is just $E(e^{c'X})$ with $X \sim N(\mu_b, Q_b)$. This is equal to $exp(c'\mu_b + \frac{1}{2}c'Q_b c)$.[31]

3.2.3 A One-Factor Model with Mixture Innovations

For the one-factor model analyzed above, the normality of the innovations carries over to yields which are also normally distributed. This implies that any linear combination of yields, in particular term spreads and first differences, are also normally distributed. The stylized facts, however, point towards the possibility that especially changes of short-term yields may not be normally distributed. Particularly, this is suggested by a high excess kurtosis, a property not compatible with the assumption of normally distributed innovations.

A simple model outlined by [11] replaces the Gaussian distribution in (3.22) by a mixture of two normal densities. This specification implies positive kurtosis in innovations that carries over to yields and yield changes. Next, we analyze the model in some detail. Our exposition also contains a little extension that allows for non-vanishing skewness.

The model consists of the factor process (3.20) and the SDF specification (3.21). The state innovation is now assumed to be distributed as a mixture of two normals,

$$u_t \sim i.i.d.\ \omega_1 N(\mu_1,\, \sigma_1^2) + \omega_2 N(\mu_2,\, \sigma_2^2) \tag{3.47}$$

$$\begin{aligned} \text{with} \quad & \omega_1\mu_1 + \omega_2\mu_2 = 0, \\ & \omega_1, \omega_2 \geq 0, \quad \omega_1 + \omega_2 = 1. \end{aligned}$$

[31] See lemma A.1 in the appendix.

From the results in section 3.2.2, the first four central moments of the distribution of u_t are given by

$$E(u_t) = 0, \tag{3.48}$$

$$E(u_t^2) = \sum_{b=1}^{2} \omega_b[\sigma_b^2 + \mu_b^2] =: \sigma^2, \tag{3.49}$$

$$E(u_t^3) = \sum_{b=1}^{2} \omega_b[3\mu_b\sigma_b^2 + \mu_b^3], \tag{3.50}$$

$$E(u_t^4) = \sum_{b=1}^{2} \omega_b[3\sigma_b^4 + 6\mu_b^2\sigma_b^2 + \mu_b^4]. \tag{3.51}$$

Consider first the case with $\mu_1 = \mu_2 = 0$. We interpret the model in such a way that u_t is drawn from a normal distribution with moderate variance σ_2^2 with a high probability ω_2, while with a small probability ω_1 the innovation is drawn from a normal distribution with large variance σ_1^2. That is, we assume

$$\omega_1 < \omega_2, \quad \sigma_1^2 > \sigma_2^2. \tag{3.52}$$

As we have seen in section 3.2.2, choosing unequal variances induces excess kurtosis of the distribution of u_t. Let c denote the ratio between the higher and the lower variance, i.e. $\sigma_1^2 = c\sigma_2^2$, $c > 1$. Then the kurtosis of u_t is strictly increasing in c as proved in the appendix.

From a dynamic viewpoint, the factor process with the mixture innovation can be interpreted as exhibiting occasional 'jumps'. Heuristically speaking, most of the time the process fluctuates moderately around its long run mean, but in $\omega_1 \cdot 100$ percent of the time, the process is highly probable to exhibit a large deviation which can be upward or downward.

By allowing (additionally) that $\mu_1 \neq \mu_2$, a skewed distribution of u_t can be established. It is derived in the appendix that the third moment of u_t can be written as

$$E(u_t^3) = \omega_1\mu_1 \left[3(\sigma_1^2 - \sigma_2^2) + \frac{\omega_2 - \omega_1}{\omega_2^2}\mu_1^2\right]. \tag{3.53}$$

Thus, under assumption (3.52), the distribution of u_t is skewed to the left if $\mu_1 < 0$ and skewed to the right if $\mu_1 > 0$.

For the Gaussian one-factor model above, yields turned out to be particularly simple functions of the state variable. The following result states that this simple structure carries over to the mixture model: under the condition of no arbitrage, yields are still affine functions of the state variable.

Proposition 3.3 (Yields in the discrete -time one-factor model with mixture innovations). *For the one-factor model (3.20) - (3.21), (3.47) zero bond yields are given as*

$$y_t^n = \frac{A_n}{n} + \frac{B_n}{n} X_t \tag{3.54}$$

with[32]

$$B_n = \sum_{i=0}^{n-1} \kappa^i = \frac{1-\kappa^n}{1-\kappa} \tag{3.55}$$

$$A_n = \sum_{i=0}^{n-1} G(B_i) \tag{3.56}$$

where

$$G(B_i) = \delta + B_i\theta(1-\kappa) - \ln\left(\sum_{b=1}^{B} \omega_b e^{-(\lambda+B_i)\mu_b + \frac{1}{2}(\lambda+B_i)^2\sigma_b^2}\right)$$

In the next section, we will consider the affine multifactor mixture model of which the one-factor mixture model is a special case. Thus, the proof given there also holds for this proposition here[33] and we omit proving the special case.

Again, the intercept parameter δ can be chosen in such a way that the state variable X_t equals the short rate. We must have $A_1 = 0$ which implies

$$\delta = \ln\left(\sum_{b=1}^{B} \omega_b e^{-\lambda\mu_b + \frac{1}{2}\lambda^2\sigma_b^2}\right). \tag{3.57}$$

The mean yield curve resulting from the mixture model is given by

$$\begin{aligned} E(y_t^n) &= \frac{A_n}{n} + \frac{B_n}{n}\theta \\ &= \delta + \theta - \frac{1}{n}\sum_{i=0}^{n-1} \ln\left(\sum_{b=1}^{2} \omega_b e^{-(\lambda+B_i)\mu_b + \frac{1}{2}(\lambda+B_i)^2\sigma_b^2}\right) \end{aligned} \tag{3.58}$$

For the Sharpe ratio, i.e. the expected excess return divided by its standard deviation, one obtains[34]

$$\begin{aligned} &SR_t^n \\ &= \frac{1}{B_{n-1}\sigma}\left[\ln\left(\sum_{b=1}^{2} \omega_b e^{\lambda\mu_b + \frac{1}{2}\lambda^2\sigma_1^2}\right) - \ln\left(\sum_{b=1}^{2} \omega_b e^{(\lambda+B_{n-1})\mu_b \frac{1}{2}(\lambda+B_{n-1})^2\sigma_b^2}\right)\right]. \end{aligned} \tag{3.59}$$

32 Empty sums are evaluated as zero.

33 In terms of the general model below one has to set $d = 1$ and $B = 2$ in order to obtain this model. The parameter a there corresponds to $\theta(1-\kappa)$ in the model here.

34 See section C.1 in the appendix.

For the plain Vasiček model, the Sharpe ratio has been shown to be an affine function of λ and to be monotonically decreasing in that parameter. Here, the Sharpe ratio is a nonlinear function of λ. However, it is shown in the appendix (for the case $\mu_1 = \mu_2 = 0$) that again the Sharpe ratio is monotonic increasing in $-\lambda$.

3.2.4 Comparison of the One-Factor Models

It is easily observed that the plain Vasiček model and the one-factor mixture model coincide if either $\sigma_1^2 = \sigma_2^2 = \sigma^2$ and $\mu_1 = \mu_2 = 0$, or if one of the ω_i is zero and the variance corresponding to the other component equals σ^2. In these cases, yields are equal for both models for all times t and for all realizations of the state variable X_t. If the mixture model exhibits a 'true' mixture distribution, the two models give rise to different term structures. In order to make a sensible comparison we assume that the parameters θ and κ are the same for the two models. It is further assumed that the variance σ^2 of the Vasiček model equals the variance implied by the mixture distribution of the other model. In the following, a tilde on top of a symbol denotes that it belongs to the plain model as opposed to the mixture model. Again, we focus on excess kurtosis and set $\mu_1 = \mu_2 = 0$. Of course, a comparison similar to the following, can also be conducted to assess the impact of different means of the component densities.

By assuming that the models are characterized by the same value for the parameter κ, it follows that the two models show the same sensitivity of yields to changes in the factor,[35]

$$\frac{\partial \tilde{y}_t^n}{\partial r_t} = \frac{\partial y_t^n}{\partial r_t} = \frac{B_n}{n} = \frac{1}{n}\frac{1-\kappa^n}{1-\kappa}.$$

If we consider yields in levels, however, the two models do differ from each other. For a given realization of the short rate, say $\bar{r}$, the difference in yields is given by

$$\begin{aligned}
D_t^n &= y_t^n - \tilde{y}_t^n \\
&= \frac{A_n}{n} + \frac{B_n}{n}\bar{r} - \frac{\tilde{A}_n}{n} - \frac{\tilde{B}_n}{n}\bar{r} \\
&= \frac{1}{n}(A_n - \tilde{A}_n) \\
&= \frac{1}{n}\sum_{i=1}^{n-1}\left(\ln\left[\sum_{b=1}^{2}\omega_b e^{\frac{1}{2}\lambda^2\sigma_b^2}\right] - \ln\left[\sum_{b=1}^{2}\omega_b e^{\frac{1}{2}(\lambda+B_i)^2\sigma_b^2}\right]\right. \\
&\quad \left. -\frac{1}{2}\tilde{\lambda}^2\tilde{\sigma}^2 + \frac{1}{2}(\tilde{\lambda}+B_i)^2\tilde{\sigma}^2\right).
\end{aligned} \tag{3.60}$$

[35] We write r_t instead of X_t to emphasize the special interpretation of X_t as the short-term interest rate, which is induced by setting δ equal to (3.31) or (3.57), respectively.

Setting $\bar{r} = \theta$ it is obvious that the latter expression also denotes the expected difference between the two term structures. This difference can be interpreted as the expected error that is induced if the true mixture distribution of the short rate innovation is falsely assumed to be a simple normal. From the difference in yields one can obtain the percentage error in zero bond prices using (2.2). We have

$$\left(\frac{\tilde{P}_t^n - P_t^n}{P_t^n}\right) \cdot 100 = \left(e^{n(y_t^n - \tilde{y}_t^n)} - 1\right) \cdot 100. \tag{3.61}$$

It can be shown that the market price of risk parameter $\tilde{\lambda}$ in the plain model cannot be chosen such that the average term structure coincides with the true (i.e. the mixture model's) term structure for all n. The extent to which the term structure implied by the plain model deviates from the true one depends on the value chosen for $\tilde{\lambda}$ and on the other model parameters, especially the variance ratio in the mixture distribution.

To illustrate the point we give a numerical example using a parameterization that is based on [11]. We set[36]

$$\begin{aligned}
\kappa &= \tilde{\kappa} = 0.976 \\
\sigma_1^2 &= a \cdot 0.001637^2, \quad a \in \{1, 2, 5, 10\}. \\
\sigma_2^2 &= 0.0004290^2 \\
\omega &= 0.05 \\
\tilde{\sigma}^2 &= \omega\sigma_1^2 + (1-\omega)\sigma_2^2 \\
\lambda &= -178.4;
\end{aligned}$$

A higher value for a implies a higher variance ratio in the mixture, i.e. a more accentuated distinction over the model with the simple normal distribution.

For each value of a, the parameter $\tilde{\lambda}$ is chosen in such a way that the one-year yield implied by the Gaussian model matches the one-year yield implied by the mixture model.[37] The left graph in figure 3.1 shows the percentage pricing error (3.61) for maturities of 1 to 240 months. The lines with higher pricing error correspond to higher values of a. The picture shows that for variance ratios of σ_1^2/σ_2^2 equal to 14.6 or 29.1 ($a = 1$ and $a = 2$) the two lines of percentage errors are hardly distinguishable from the zero line. That is there is no pronounced difference between the two models. For the variance ratios of 72.8 and 145.7 ($a = 5$ and $a = 10$) the difference turns out to be quite substantial.

The right picture has a similar content, but this time $\tilde{\lambda}$ is chosen in order to equate the ten-year yields of the two models.

[36] Note that the difference in yields and prices is not dependent on θ.
[37] This is done by numerically solving $y_t^n - \tilde{y}_t^n = 0$ for $n = 12$ with respect to $\tilde{\lambda}$.

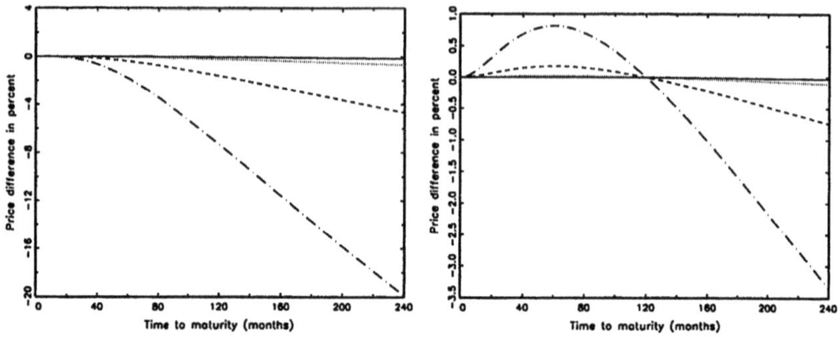

Fig. 3.1. Percentage errors of zero bond prices when using the Vasiček model instead of the mixture model. The short-dashed line corresponds to $a = 5$, the dashed-dotted line to $a = 10$. Lines for $a = 1$ and $a = 2$ are hardly distinguishable from the zero line.

We have demonstrated that the two models exhibit a different mean yield curve. In the next subsection, higher moments of the unconditional distributions of bond yields are computed, and the results for the two models are compared.

3.2.5 Moments of the One-Factor Models

Our factor process is an AR(1) which is second-order stationary, i.e. its mean and all autocovariances are finite and time invariant. In section B in the appendix we prove that the third and fourth moments of the process are also time invariant and finite.[38] It has already been computed above, that the expectation of the y_t^n, i.e. the mean yield, is given by $A_n/n + B_n/n \cdot \theta$, where θ is the mean of the state process. In the following, variables with an asterisk denote deviations from the mean:

$$X_t^* = X_t - \theta, \quad y_t^{n*} = y_t^n - E(y_t^n) = \frac{B_n}{n}(X_t - \theta) = \frac{B_n}{n} X_t^*.$$

For both one-factor models, higher moments are given as

$$E[(y_t^{n*})^i] = \left(\frac{B_n}{n}\right)^i E[(X_t^*)^i], \quad i = 2, 3, 4.$$

Thus, the moments of yields are proportional to those of the state variable. For the latter one obtains

[38] This is done for the more general case of a vector AR(1) process, of which the scalar process considered in this section is of course a special case.

$$E[(X_t^*)^2] = \frac{\sigma^2}{1-\kappa^2},$$
$$E[(X_t^*)^3] = \frac{E(u_t^3)}{1-\kappa^3}.$$
$$E[(X_t^*)^4] = \frac{6\kappa^2\sigma^4 + (1-\kappa^2)E(u_t^4)}{(1-\kappa^2)(1-\kappa^4)}.$$

These expressions have been derived making use of the stationarity assumption. For the detailed computations see the appendix.

We have seen above that the mean yield curves of the two models differ from each other due to the different functional forms of the intercept A_n. Now we observe that if both models possess the same innovation variance σ^2, the variance of bond yields,

$$Var(y_t) = E[(y_t^{n*})^2] = \left(\frac{B_n}{n}\right)^2 \frac{\sigma^2}{1-\kappa^2} \tag{3.62}$$

will be the same for both models. For the covariance, one obtains

$$Cov(y_t^m, y_t^n) = E\left[\left(\frac{B_m}{m}X_t^*\right)\left(\frac{B_n}{n}X_t^*\right)\right] = \frac{B_m}{m}\frac{B_n}{n}E(X_t^{*2}), \tag{3.63}$$

which implies that any two bond yields are perfectly correlated for each time t, i.e.

$$Corr(y_t^m, y_t^n) = 1. \tag{3.64}$$

This is a direct consequence of the following two key characteristics that are shared by the plain and the mixture model: first, there is only one source of randomness that drives the whole term structure; second, all yields are affine functions of the factors. However, perfect correlation of bond yields is not consistent with the stylized facts. Bond yields do show highly positive contemporaneous correlation, but it is not perfect and it is different for different pairs of maturities.

For the Gaussian case, $E(u_t^3) = 0$, thus $E(X_t^{*3}) = 0$ and the coefficient of skweness is zero at all maturities. For the case of a normal mixture, the situation $\mu_1 \neq \mu_2$ leads to $E(X_t^{*3}) \neq 0$ which induces a skewed distribution of bond yields. The coefficient of skewness is proportional to the skewness of the innovation and will be the same for all maturities. We have

$$\begin{aligned} skew(y_t^n) &= \frac{E[(y_t^{n*})^3]}{(E[(y_t^{n*})^2])^{\frac{3}{2}}} \\ &= \frac{\left(\frac{B_n}{n}\right)^3 E[(X_t^{n*})^3]}{\left(\frac{B_n}{n}\right)^{2\frac{3}{2}} (E[(X_t^{n*})^2])^{\frac{3}{2}}} \\ &= \frac{(1-\kappa^2)^{\frac{3}{2}}}{1-\kappa^3} skew(u_t), \end{aligned}$$

i.e. the maturity dependent terms cancel from the expression. If $\mu_1 = \mu_2 = 0$, then the mixture model also leads to symmetric distributions of bond yields.

Similar results hold for fourth moments. For yields, the coefficient of kurtosis can be written in terms of the kurtosis of the state innovation:[39]

$$\begin{aligned} kurt(y_t^n) &= \frac{E[(y_t^{n*})^4]}{(E[(y_t^{n*})^2])^2} - 3 \\ &= \frac{\left(\frac{B_n}{n}\right)^4 E[(X_t^{n*})^4]}{\left(\frac{B_n}{n}\right)^4 (E[(X_t^{n*})^2])^2} - 3 \\ &= \frac{(1-\kappa)^2}{1+\kappa^2} kurt(u_t). \end{aligned}$$

Clearly, for the Gaussian model, the kurtosis is zero.[40] For the mixture model, excess kurtosis in the state innovation induces excess kurtosis in bond yields. The kurtosis of bond yields rises proportionally to the kurtosis in the innovation term, which in turn increases in the ratio of the two component variances, σ_1^2 and σ_2^2, as commented above. By taking the derivative with respect to κ one can also show that an increase in the autoregression parameter leads to a decrease in kurtosis. Finally, the kurtosis of bond yields is independent of time to maturity n.

For first differences of yields, i.e. for

$$\Delta y_t^n = y_t^n - y_{t-1}^n = \frac{B_n}{n}(\Delta X_t)$$

the first two moments are given as[41]

$$E(\Delta y_t^n) = 0, \quad \text{and} \quad Var(\Delta y_t^n) = E[(\Delta y_t^n)^2] = \left(\frac{B_n}{n}\right)^2 \frac{2}{1+\kappa^2}\sigma^2. \tag{3.65}$$

This implies that for $\kappa > 0.5$ yield changes have a smaller variance than yield levels. By similar computations as above, changes of yields also exhibit perfect contemporaneous correlation,

$$Corr(\Delta y_t^n, \Delta y_t^m) = 1.$$

For skewness and kurtosis we obtain

$$skew(\Delta y_t^n) = \frac{3\kappa(1-\kappa)(1+\kappa)^{\frac{3}{2}}}{(1-\kappa^3)2^{\frac{3}{2}}} skew(u_t)$$

and

---

39 See the appendix for the computation.

40 This follows already from the fact that the stationary distribution of y_t^n is Gaussian.

41 Again, moments for differenced yields are computed in the appendix.

$$kurt(\Delta y_t^n) = \frac{(1-\kappa)^4 + (1+\kappa^2)(1+\kappa)^2}{4(1+\kappa^2)} kurt(u_t)$$

respectively. Again, both measures do not depend on time to maturity and are proportional to the corresponding quantity of u_t.

Against the background of the stylized facts, it is also instructive to compare the higher moments of yields in levels and yields in first differences. It turns out that both the coefficient of skewness (in absolute terms) and the coefficient of kurtosis are higher for yield changes than for levels if κ is sufficiently high. Formally,

$$|skew(\Delta y_t^n)| > |skew(y_t^n)|, \quad \text{for } 0.598 < \kappa < 1$$

and

$$kurt(\Delta y_t^n) > kurt(y_t^n), \quad \text{for } 0.285 < \kappa < 1.$$

Summing up, the mixture model is able to capture skewness and excess kurtosis in bond yields. Concerning kurtosis, it also implies that kurtosis for yield changes exceeds that for levels. However, the one factor model does not allow for kurtosis that changes over the range of maturities.

Finally, we compute the autocorrelation of yields and yield changes. For both levels and differences, the coefficient of autocorrelation is identical to that of the single factor. We have

$$Corr(y_t^n, y_{t-1}^n) = \frac{E\left(\frac{B_n}{n} X_t^* \frac{B_n}{n} X_{t-1}^*\right)}{E\left[\left(\frac{B_n}{n} X_t^*\right)^2\right]} = \kappa \tag{3.66}$$

and

$$Corr(\Delta y_t^n, \Delta y_{t-1}^n) = \frac{E\left(\frac{B_n}{n} \Delta X_t \frac{B_n}{n} \Delta X_{t-1}\right)}{E\left[\left(\frac{B_n}{n} \Delta X_t\right)^2\right]} = -\frac{1}{2}(1-\kappa). \tag{3.67}$$

Yields of all maturities show the same autocorrelation which is not consistent with the stylized facts. If autocorrelation is high for levels, then the autocorrelation of first differences is low. This feature is consistent with the data. However, the model implies that for positive autocorrelation in levels, there is negative autocorrelation in first differences.

3.3 Affine Multifactor Gaussian Mixture Models

It is one of the major drawbacks of one-factor models that they are not flexible enough to simultaneously fit the cross-section pattern and the dynamics of the yield curve.[42] If κ is chosen such that the model captures the high autocorrelation of yields, the model is typically not able to capture the curvature of the

[42] See, e.g., [11] and [26].

mean yield curve. This is rooted in the fact that the parameter κ shows up in the factor process and, via the condition of no arbitrage, in the cross-sectional parameters A_n and B_n.

The flexibility of the model to adapt to both the dynamics and the cross-sectional pattern of the term structure can be increased by introducing more factors. In the following we derive the yield equations for the general case of a Gaussian model with d factors. After that we introduce the class of affine multifactor Gaussian mixture (AMGM) models. This class allows for an arbitrary number of factors as well as an arbitrary number of components in the mixture distribution. For models of this class, we introduce a canonical representation.

3.3.1 Model Structure and Derivation of Arbitrage-Free Yields

Let X_t denote a d-dimensional vector of factors. Its dynamic evolution is specified as a VAR(1) process, i.e.

$$X_t = a + \mathcal{K}X_{t-1} + u_t, \tag{3.68}$$

where a is a $d \times 1$ vector of constants, and $\mathcal{K}$ is a $d \times d$ matrix. The eigenvalues of $\mathcal{K}$ are assumed to lie inside the unit circle, which guarantees stationarity of the process $\{X_t\}$. The $d \times 1$ innovation vector u_t has a multivariate normal distribution and is serially independent,

$$u_t \sim i.i.d.\, N(0, V). \tag{3.69}$$

The pricing kernel is affine in the vector of factors and its innovations,

$$-\ln M_{t+1} = \delta + \gamma' X_t + \lambda' u_{t+1}, \tag{3.70}$$

where δ is a scalar and γ and λ are both $d \times 1$ vectors. As for the one-factor models, the components of the vector λ will be referred to as the market price of risk parameters.

Proceeding in a similar fashion as for the one-factor models one obtains that bond yields are affine in X_t.

Proposition 3.4 (Yields in the linear Gaussian multifactor model). *For the linear Gaussian multifactor model (3.68) - (3.70) zero bond yields are given as*

$$y_t^n = \frac{A_n}{n} + \frac{1}{n} B_n' X_t \tag{3.71}$$

with[43]

$$B_n = (I - \mathcal{K}'^n)(I - \mathcal{K}')^{-1}\gamma \tag{3.72}$$

$$A_n = \sum_{i=0}^{n-1} G(B_i) \tag{3.73}$$

43 Empty sums are evaluated as zero.

where

$$G(B_i) = \delta + B_i'a - \frac{1}{2}(\lambda + B_i)'V(\lambda + B_i).$$

Proof. We start with the guess that bond prices are affine in factors,

$$-\ln P_t^n = A_n + B_n'X_t$$

and use the fundamental pricing equation

$$-\ln P_t^{n+1} = -\ln E_t(M_{t+1}P_{t+1}^n)$$

for the computation of the functional form of the scalar A_n and the d-dimensional vector B_n. The expression

$$\begin{aligned}
&\ln M_{t+1} + \ln P_{t+1}^n \\
&= -\delta - \gamma'X_t - \lambda'u_{t+1} - A_n - B_n'X_{t+1} \\
&= -\delta - \gamma'X_t - \lambda'u_{t+1} - A_n - B_n'(a + \mathcal{K}X_t + u_{t+1}) \\
&= -\delta - A_n - B_n'a - (\gamma' + B_n'\mathcal{K})X_t - (\lambda' + B_n')u_{t+1} \\
&=: V_{t+1}
\end{aligned}$$

is normally distributed conditional on $\mathcal{F}_t$ with conditional mean and variance-covariance matrix given by

$$E_t(V_{t+1}) = -\delta - A_n - B_n'a - (\gamma' + B_n'\mathcal{K})X_t$$

and

$$Var_t(V_{t+1}) = (\lambda + B_n)'V(\lambda + B_n)$$

respectively. Thus, the conditional expectation on the right hand side of the pricing equation is

$$\begin{aligned}
&-\ln E_t(M_{t+1}P_{t+1}^n) \\
&= -\ln E_t\left(e^{\ln M_{t+1} + \ln P_{t+1}^n}\right) \\
&= \delta + A_n + B_n'a + (\gamma' + B_n'\mathcal{K})X_t - \frac{1}{2}(\lambda + B_n)'V(\lambda + B_n)
\end{aligned}$$

Equating this expression with our guess for log bond prices yields the difference equations:[44]

$$B_{n+1} = \gamma + \mathcal{K}'B_n \tag{3.74}$$

$$A_{n+1} = A_n + \delta + B_n'a - \frac{1}{2}(\lambda + B_n)'V(\lambda + B_n) \tag{3.75}$$

with initial conditions $A_0 = 0$ and $B_0 = 0$. Solving the difference equation for B_n, one obtains

[44] Note that A_n is a scalar whereas B_n is a $d \times 1$ vector.

$$B_n = (I + \mathcal{K}' + \mathcal{K}'^2 + \ldots + \mathcal{K}'^{n-1})\gamma \tag{3.76}$$

which yields

$$B_n = (I - \mathcal{K}'^n)(I - \mathcal{K}')^{-1}\gamma. \tag{3.77}$$

Having B_n at hand, formula (3.73) for A_n follows by solving the difference equation (3.75). □

In analogy to the one-factor model above, we introduce nonnormality of the innovation vector by specifying its distribution as a mixture of B normal distributions. That is we consider the model (3.68), (3.70) with

$$u_t \sim i.i.d. \sum_{b=1}^{B} \omega_b N(\mu_b, V_b), \quad \sum_{b=1}^{B} \omega_b = 1, \ \sum_{b=1}^{B} \omega_b \mu_b = 0. \tag{3.78}$$

Again, the affine structure is maintained when moving from a d-variate normal distribution to a d-variate mixture, as implied by the following proposition.

Proposition 3.5 (Yields in the linear multifactor Gaussian mixture model). *For the multifactor model (3.68), (3.70), (3.78), zero bond yields are given as*

$$y_t^n = \frac{A_n}{n} + \frac{1}{n} B_n' X_t \tag{3.79}$$

with[45]

$$B_n = (I - \mathcal{K}'^n)(I - \mathcal{K}')^{-1}\gamma \tag{3.80}$$

$$A_n = \sum_{i=0}^{n-1} G(B_i) \tag{3.81}$$

where

$$G(B_i) = \delta + B_i' a - \ln \left[\sum_{b=1}^{B} \omega_b \cdot e^{-(\lambda + B_i)'\mu_b + \frac{1}{2}(\lambda + B_i)' V_b (\lambda + B_i)} \right].$$

Proof. The proof has the same structure as the one for the Gaussian multifactor model. Again we have

$$\begin{aligned} & \ln M_{t+1} + \ln P_{t+1}^n \\ &= -\delta - A_n - B_n' a - (\gamma' + B_n' \mathcal{K}) X_t - (\lambda' + B_n') u_{t+1} \\ &:= V_{t+1} \end{aligned}$$

This time, however, the conditional distribution of V_{t+1} is not normal but a d-variate normal mixture with B components. We have to compute

$$E_t \left(e^{\ln M_{t+1} + \ln P_{t+1}^n} \right)$$

[45] Empty sums are evaluated as zero.

which has the form

$$E_t\left(e^{c_0+c_1'u_{t+1}}\right)$$

with $c_0 = -\delta - A_n - B_n'a - (\gamma' + B_n'\mathcal{K})X_t$, $c_1 = -(\lambda + B_n)$. Using the result (3.46) from above we have

$$\begin{aligned} & E_t\left(e^{c_0+c_1'u_{t+1}}\right) \\ & = e^{c_0}\left(\sum_{b=1}^{B} \omega_b\, e^{c_1'\mu_b+\frac{1}{2}c_1'V_b c_1}\right). \end{aligned}$$

Plugging back in the original variables we thus obtain

$$\begin{aligned} & \ln E_t\left(e^{\ln M_{t+1}+\ln P_{t+1}^n}\right) \\ & = -\delta - A_n - B_n'a - (\gamma' + B_n'\mathcal{K})X_t \\ & \quad + \ln\left[\sum_{b=1}^{B} \omega_b \cdot e^{-(\lambda+B_n)'\mu_b+\frac{1}{2}(\lambda+B_n)'V_b(\lambda+B_n)}\right] \end{aligned}$$

For the fundamental pricing equation (3.23) to hold, the coefficient functions A_n and B_n have to satisfy the following set of difference equations

$$B_{n+1} = \gamma + \mathcal{K}'B_n \tag{3.82}$$

$$\begin{aligned} A_{n+1} &= \delta + A_n + B_n'a \\ & \quad - \ln\left[\sum_{b=1}^{B} \omega_b \cdot e^{-(\lambda+B_n)'\mu_b+\frac{1}{2}(\lambda+B_n)'V_b(\lambda+B_n)}\right]. \end{aligned} \tag{3.83}$$

with initial conditions $A_0 = 0$ and $B_0 = 0$. The vector difference equation for B_n is the same as in the Gaussian multifactor model, so again

$$B_n = (I + \mathcal{K}' + \mathcal{K}'^2 + \ldots + \mathcal{K}'^{n-1})\gamma = (I - \mathcal{K}'^n)(I - \mathcal{K}')^{-1}\gamma.$$

The solution of the difference equation for A_n leads to (3.81). □

The last specification is the most general considered so far as it nests both one-factor models and the multifactor Gaussian model as special cases.

We introduce the short hand notation

$$\mathcal{M}_{Mixt}(\mathcal{K}, a, \{V_b\}, \{\mu_b\}, \{\omega_b\}, \lambda, \gamma, \delta;\ d, B)$$

which represents the set of equations (3.68), (3.78), (3.79) together with specific numerical values for the model parameters, a specific dimension d of the factor vector and a specific number B of mixture components. We refer to $\mathcal{M}_{Mixt}(\cdot)$ as a *model structure*. With $\{V_b\}$ we denote $\{V_1, \ldots, V_B\}$, the set of component variance-covariance matrices. $\{\mu_b\}$ and $\{\omega_b\}$ have an analogous meaning. The set of all such model structures will be referred to as the *class of affine multifactor Gaussian mixture (AMGM) models.*

Concerning related literature, the class of 'affine models' by Duffie and Kan [45][46] is characterized by a factor process of the form

$$X_t = a + \mathcal{K}X_{t-1} + [V(X_{t-1})]^{\frac{1}{2}}\, u_t$$

where

$$u_t \sim i.i.d.N(0, I_d)$$

and $V(X_{t-1})$ is a diagonal matrix with ith diagonal element given by

$$V(X_{t-1}) = \alpha_i + \beta_i' X_{t-1}.$$

The pricing kernel satisfies

$$-\ln M_{t+1} = \delta + \gamma' X_t + \lambda'[V(X_t)]^{\frac{1}{2}}\, u_{t+1}.$$

Also for this model, yields are affine functions of the factors. The model is able to capture level-dependent volatility, i.e the conditional variance of the factors depends on the previous period's factor realization. For $d = 1$, i.e. one factor only, the model is the discrete-time analogy of the famous model by Cox, Ingersoll and Ross [34].[47]

Homoscedastic Gaussian models are contained in both classes: models from Duffie and Kan's class are Gaussian if $\beta_i = 0$ for all i; models from the AMGM class are Gaussian if the mixture distribution contains $B = 1$ component only. However, genuine mixture models are not nested within the class of [45]; conversely, models with state dependent volatility are not nested within the AMGM class.

Another strand of literature deals with models containing regime changes. Our mixture models can be interpreted as special cases of those regime-switching models. For the mixture models in this book, in each period the innovation is drawn from B different regimes, each regime being characterized by its own normal distribution $N(\mu_b, V_b)$. The probability ω_b of drawing from a particular regime, however, is the same in every period and it is independent of the past. In contrast, the regime switching models by [13] and [36][48] let additional model parameters (i.e. besides the innovation variance) depend on the prevailing regime. Moreover, the regime follows a Markov-switching process. Yields in these models are still affine functions of the state variables.

3.3.2 Canonical Representation

For specific numerical values of the parameters, the multifactor mixture model describes the dynamics of the term structure of interest rates: for an arbitrary

[46] See [11] for the discrete-time version.
[47] See [104] for the discrete version.
[48] See also the references given therein.

collection of maturities $\{n_1, \ldots, n_k\}$, the evolution of $(y_t^{n_1}, \ldots, y_t^{n_k})'$ is determined by the factor process and the relation between factors and yields. However, we will now show that two different parameterizations of the mixture model may give rise to the same term structure.[49]

Let $\mathcal{M}_0 = \mathcal{M}_{Mixt}(\mathcal{K}, a, \{V_b\}, \{\mu_b\}, \{\omega_b\}, \lambda, \gamma, \delta;\, d, B)$ be a model structure, and let X_t denote its corresponding factor vector. We define an *invariant affine transformation* $\mathcal{S}$ of $\mathcal{M}_0$ by a $d \times 1$ vector l and an invertible $d \times d$ matrix L, such that the factor vector of the transformed model is given by

$$X_t^L := LX_t + l, \tag{3.84}$$

and its model parameters are given by

$$\begin{aligned} a^L &= l + La - L\mathcal{K}L^{-1}l, \quad \mathcal{K}^L = L\mathcal{K}L^{-1}, \\ \mu_b^L &= L\mu_b, \quad V_b^L = LVL', \quad \omega_b^L = \omega_b, \end{aligned} \tag{3.85}$$

and

$$\delta^L = \delta - \gamma' L^{-1} l, \quad \gamma^L = L^{-1'}\gamma, \quad \lambda^L = L^{-1'}\lambda. \tag{3.86}$$

For the new model structure $\mathcal{M}_{Mixt}^L$ arising from the transformation we write $\mathcal{M}_{Mixt}^L = \mathcal{S}(\mathcal{M}_0; L, l)$. The next result shows that the invariant transformation deserves its name.[50] We prove that the new factor process is also governed by a VAR(1) whose innovation distribution is again a B-component mixture, and that the transformed model implies the same term structure.

Proposition 3.6 (Invariant transformation of the multifactor mixture model). *Let y_t^n denote the yields implied by the model structure $\mathcal{M}_0 = \mathcal{M}_{Mixt}(\mathcal{K}, a, \{V_b\}, \{\mu_b\}, \{\omega_b\}, \lambda, \gamma, \delta;\, d, B)$ and denote its factor process by $\{X_t\}$. Define a new factor vector X_t^L by (3.84). Then:*

i) The new factor process $\{X_t^L\}$ satisfies:

$$X_t^L = a^L + \mathcal{K}^L X_{t-1}^L + u_t^L, \quad u_t^L \sim \sum_{b=1}^{B} \omega_b^L N(\mu_b^L, V_b^L) \tag{3.87}$$

with parameters given by (3.85).

ii) Denote by y_t^{Ln} the yields implied by the model with factor process (3.87) and pricing kernel equation

$$-\ln M_{t+1}^L = \delta^L + \gamma^{L'} X_t^L + \lambda^{L'} u_{t+1}^L, \tag{3.88}$$

with parameters given by (3.86). Then $y_t^n = y_t^{Ln}$ for all t and n.

---

[49] Of course, the argument also holds true for the Gaussian model as a special case.

[50] Compare the invariant transformations and canonical representations for the class of continuous-time exponential-affine models discussed by [35].

Proof. The old factor vector in terms of the new one is given by

$$X_t = L^{-1}X_t^L - L^{-1}l.$$

This is plugged into the original VAR(1) process

$$X_t = a + \mathcal{K}X_{t-1} + u_t$$

from model $\mathcal{M}^0_{Mixt}$ above. We have

$$L^{-1}X_t^L - L^{-1}l = a + \mathcal{K}L^{-1}X_{t-1}^L - \mathcal{K}L^{-1}l + u_t.$$

Premultiplying the equation by L and rearranging yields

$$X_t^L = (l + La - L\mathcal{K}L^{-1}l) + (L\mathcal{K}L^{-1})X_{t-1}^L + Lu_t.$$

Making use of the result in section 3.2.2, the distribution of $u_t^L := Lu_t$ is again a B-component mixture,

$$u_t^L \sim \sum_{b=1}^{B} \omega_b N(L\mu_b,\, LV_bL'),$$

which completes the proof of *i)*.

For *ii)* we have to show that yields obtained from $\mathcal{M}^L_{Mixt}$ are equal to those from $\mathcal{M}^0_{Mixt}$.[51] That is, $y_t^{Ln} = y_t^n$ or

$$A_n^L + B_n^{L\prime}X_t^L = A_n + B_n'X_t. \tag{3.89}$$

The coefficient vector B_n^L satisfies (3.76).[52] So we have

$$\begin{aligned} B_n^L &= \left(I + \mathcal{K}^{L\prime} + (\mathcal{K}^{L\prime})^2 + \ldots + (\mathcal{K}^{L\prime})^{n-1}\right)\gamma^L \\ &= \left(L^{-1\prime}L' + L^{-1\prime}\mathcal{K}'L' + L^{-1\prime}\mathcal{K}'^2L' + \ldots + L^{-1\prime}\mathcal{K}'^{n-1}L'\right)L^{-1\prime}\gamma \\ &= L^{-1\prime}\left(I + \mathcal{K} + \mathcal{K}'^2 + \ldots + \mathcal{K}'^{n-1}\right)\gamma \\ &= L^{-1\prime}B_n. \end{aligned}$$

Thus, for the second addend of the left hand side of (3.89) we have

$$\begin{aligned} B_n^{L\prime}X_t^L &= B_n'L^{-1}LX_t + B_n'L^{-1}l \\ &= B_n'X_t + B_n'L^{-1}l. \end{aligned}$$

---

51 We prove this by using the explicit term structure equation, i.e. yields as a function of the factors. Taking this route, we also obtain explicit expressions of the coefficients A_n^L and B_n^L for the transformed model. However, the assertion can be proved simply by showing that $\ln M_t = \ln M_t^L$. Then the prices of the transformed model must solve the same fundamental pricing equation (3.10) as those of the original model and will be the same, accordingly.

52 That formula for B_n holds for the Gaussian model and the mixture model.

It remains to be shown that

$$A_n^L = A_n - B_n' L^{-1} l. \tag{3.90}$$

First note that

$$\begin{aligned}
&(\lambda^L + B_n^L)' V_b^L (\lambda^L + B_n^L) \\
&= (\lambda + B_n)' L^{-1} L V_b L' L^{-1\prime} (\lambda + B_n) \\
&= (\lambda + B_n)' V_b (\lambda + B_n).
\end{aligned} \tag{3.91}$$

Now we prove (3.90) by induction. For $n = 1$:

$$\begin{aligned}
A_1^L &= \delta^L - \ln\left[\sum_{b=1}^{B} \omega_b^L e^{-\lambda^{L\prime}\mu_b^L + \frac{1}{2}\lambda^{L\prime} V_b^L \lambda^L}\right] \\
&= \delta - \gamma' L^{-1} l - \ln\left[\sum_{b=1}^{B} \omega_b e^{-\lambda'\mu_b + \frac{1}{2}\lambda' V_b \lambda}\right] \\
&= A_1 - B_1' L^{-1} l.
\end{aligned}$$

Assume that A_n^L satisfies (3.90) for some n. Then for $n+1$:

$$\begin{aligned}
A_{n+1}^L &= A_n^L + \delta^L + B_n^{L\prime} a^L \\
&\quad - \ln\left[\sum_{b=1}^{B} \omega_b^L \cdot e^{-(\lambda^L + B_n^L)'\mu_b^L + \frac{1}{2}(\lambda^L + B_n^L)' V_b^L (\lambda^L + B_n^L)}\right] \\
&= A_n - B_n' L^{-1} l + \delta - \gamma' L^{-1} l + B_n' L^{-1}(l + La - L\mathcal{K}L^{-1} l] \\
&\quad - \ln\left[\sum_{b=1}^{B} \omega_b \cdot e^{-(\lambda + B_n)'\mu_b + \frac{1}{2}(\lambda + B_n)' V_b (\lambda + B_n)}\right] \\
&= A_{n+1} - B_n' L^{-1} l - \gamma' L^{-1} l + B_n' L^{-1} l - B_n' \mathcal{K} L^{-1} l \\
&= A_{n+1} - (B_n' \mathcal{K} + \gamma') L^{-1} l \\
&= A_{n+1} - B_{n+1}' L^{-1} l.
\end{aligned}$$

The first equality is just the difference equation (3.83) that the sequence of coefficients $\{A_n^L\}$ has to satisfy. The second equality uses the induction assumption, expresses the $\mathcal{M}_{Mixt}^L$ parameters in terms of the $\mathcal{M}_{Mixt}^0$ parameters, and uses (3.91). The next step makes use of the difference equation (3.83) but this time for the coefficient A_n of the original model. The penultimate equality simplifies, and the last one uses the difference equation (3.82) for the coefficient vector B_n. □

We will now employ the concept of invariant affine transformations in order to reduce the number of free parameters in the multifactor models. We will justify this reduction as innocuous by starting from an arbitrarily parameterized model and applying invariant affine transformations that lead to a model structure with less free parameters.

For the multifactor mixture model, (3.68), (3.70), (3.78), we will assume that:

(C1) $a = (0, \ldots, 0)'$.
(C2) $\gamma = (1, \ldots, 1)'$.
(C3) V_1 is a diagonal matrix.
(C4) $\mathcal{K}$ is a lower triangular matrix with eigenvalues less than one in absolute value.

If a model structure satisfies properties (C1) - (C4) it is said to be in *canonical form.*

To justify these assumptions, we take an arbitrary multifactor Gaussian model structure $\mathcal{M}_0 = \mathcal{M}_{Mixt}(\mathcal{K}^0, a^0, \{V_b^0\}, \{\mu_b^0\}, \{\omega_b^0\}, \lambda^0, \gamma^0, \delta^0; d, B)$ as starting point. Here, a^0 and γ^0 are arbitrary nonzero $d \times 1$ vectors, the V_b^0 are symmetric, positive definite $d \times d$ matrices, and $\mathcal{K}^0$ is a $d \times d$ matrix with real eigenvalues that are smaller than one in absolute value. The following sequence of invariant affine transformations will lead to a model that generates exactly the same term structure and satisfies assumptions (C1) - (C4).

The first transformation diagonalizes the variance-covariance matrix of the first component. Since V_1^0 is symmetric with positive real eigenvalues, we can represent it by its spectral decomposition $V_1^0 = C\Lambda C'$ where C is the matrix of normed eigenvectors of V_1^0 and Λ is the diagonal matrix with the eigenvalues of V_1^0 on the diagonal. Moreover, C is orthogonal, i.e. $C' = C^{-1}$. We apply the affine transformation (L^1, l^1) with $L^1 = C'$ and $l^1 = 0$ to the model structure $\mathcal{M}_0$. For the transformed model structure, $\mathcal{M}_1 = \mathcal{S}(\mathcal{M}_0; L^1, l^1)$, the matrix $V_1^1 = C' V_1^0 C$ is now diagonal. The new $\mathcal{K}$ matrix, $\mathcal{K}^1 = C'\mathcal{K}^0(C')^{-1} = C'\mathcal{K}^0 C$ has the same eigenvalues as $\mathcal{K}^0$.[53]

For the next transformation, we choose $L^2 = (V_1^1)^{-0.5}$ and $l^2 = 0$. This leads to a transformed model structure with $V_1^2 = (V_1^1)^{-0.5} V_1^1 (V_1^1)^{-0.5} = I$, and $\mathcal{K}^2 = (V_1^1)^{-0.5}\mathcal{K}^1(V_1^1)^{0.5}$. Again, the new matrix $\mathcal{K}^2$ has the same eigenvalues as $\mathcal{K}^1$.

In the third step, we use the Schur decomposition of $\mathcal{K}^2$. Since $\mathcal{K}^2$ has real eigenvalues, we can write $\mathcal{K}^2 = ZSZ'$, where Z is an orthogonal matrix and S is an upper diagonal matrix, with the eigenvalues of $\mathcal{K}^2$ on the main diagonal. Choosing $L^3 = Z'$ and $l^3 = 0$ leads to $\mathcal{K}^3 = Z'\mathcal{K}^2 Z$, an upper diagonal matrix, and $V_1^3 = Z'IZ = I$.

We transform the upper triangular matrix $\mathcal{K}^3$ into a lower triangular matrix by choosing the permutation matrix

$$L^4 = \begin{pmatrix} 0 & \ldots & 0 & 1 \\ 0 & \ldots & 1 & 0 \\ \vdots & \ddots & 0 & 0 \\ 1 & \ldots & 0 & 0 \end{pmatrix}, \quad \text{and} \quad l^4 = 0$$

[53] See, e.g., [57].

in the next step. This leads to $\mathcal{K}^4 = L^4\mathcal{K}^3(L^4)^{-1}$ being lower triangular while pertaining the same eigenvalues, and $V_1^4 = L^4 I (L^4)^{-1} = I$.

We have reached the model structure

$$\mathcal{M}_4 = \mathcal{M}_{Mixt}(\mathcal{K}^4, a^4, \{V_b^4\}, \{\mu_b^4\}, \{\omega_b^4\}, \lambda^4, \gamma^4, \delta^4; d, B).$$

We have already transformed the $\mathcal{K}$ matrix to be lower triangular and the first of the variance-covariance matrices to be diagonal. The last two steps lead to the proposed forms for γ and a.

With

$$L^5 = \begin{pmatrix} \gamma_1^4 & 0 & \dots & 0 \\ 0 & \gamma_2^4 & 0 & 0 \\ \vdots & \vdots & \ddots & \vdots \\ 0 & 0 & \dots & \gamma_d^4 \end{pmatrix}, \qquad l^5 = 0,$$

where γ_i^4 denotes the ith entry of the vector γ^4, the transformed model $\mathcal{M}_5$ has $\gamma^5 = (L^5)^{-1}\gamma^4 = (1, \dots, 1)'$. The matrix V_1^5 is diagonal and $\mathcal{K}^5$ is lower-triangular with the same elements on the main diagonal as $\mathcal{K}^4$.

Finally, we make the factor evolution a mean zero process, via the transformation $L^6 = I$ and $l^6 = -(I + \mathcal{K}^5)^{-1}a^5$. Using (3.85), this leads to $a^6 = -(I + \mathcal{K}^5)^{-1}a^5 + a^5 - \mathcal{K}^5\,(I + \mathcal{K}^5)^{-1}a^5 = 0$. The only other model parameter affected by this transformation is the intercept δ from the SDF equation.

Note that in all of the steps above, the other variance-covariance matrices $V_2, \dots, V_B$ are affected by the transformations. However, they remain positive definite, since for V_b being positive definite, LV_bL' is still positive definite.

In the Gaussian model we have only one variance-covariance matrix, i.e. $V_1 = V$. Since the Gaussian model is just the mixture model with $B = 1$ component, we have just shown that V can be assumed to be diagonal without loss of generality.[54] However, it is not innocuous to assume diagonality for all component matrices of a mixture model with $B > 1$ components. This becomes obvious in light of the steps we have done for the derivation of the canonical representation: applying an invariant transformation in order to diagonalize one of the variance-covariance matrices, the remaining $B - 1$ matrices will generally lose this property.

The canonical representation implies a parsimonious parameterization of the model. In general, however, the canonical representation is not yet a unique representation for a particular model of interest. Consider, for instance, a model in which $\mathcal{K}$ and all V_b matrices are diagonal. Then an equivalent representation of the model arises from a permutation of the factors.

[54] In the canonical representation, correlation between factors is captured by the autoregressive matrix $\mathcal{K}$.

3.3.3 Moments of Yields

Similar to the one-factor models, we derive first, second, third and fourth moments for the linear multifactor model with a mixture innovation. This will be done under the specializing assumptions that the matrix $\mathcal{K}$ is diagonal, all V_b matrices are diagonal, and all the μ_b are zero. That is, we consider the multifactor model with a factor process of the form

$$\begin{pmatrix} X_{1t} \\ X_{2t} \\ \vdots \\ X_{dt} \end{pmatrix} = \begin{pmatrix} \kappa_1 & 0 & \dots & 0 \\ 0 & \kappa_2 & \dots & 0 \\ \vdots & \vdots & \ddots & \vdots \\ 0 & 0 & \dots & \kappa_d \end{pmatrix} \begin{pmatrix} X_{1\,t-1} \\ X_{2\,t-1} \\ \vdots \\ X_{d\,t-1} \end{pmatrix} + \begin{pmatrix} u_{1\,t} \\ u_{2\,t} \\ \vdots \\ u_{d\,t} \end{pmatrix} \tag{3.92}$$

with

$$\begin{pmatrix} u_{1\,t} \\ u_{2\,t} \\ \vdots \\ u_{d\,t} \end{pmatrix} \sim i.i.d. \sum_{b=1}^{B} \omega_b \, N \left(\begin{pmatrix} 0 \\ 0 \\ \vdots \\ 0 \end{pmatrix}, \begin{pmatrix} v_{1b}^2 & 0 & \dots & 0 \\ 0 & v_{2b}^2 & \dots & 0 \\ \vdots & \vdots & \ddots & \vdots \\ 0 & 0 & \dots & v_{db}^2 \end{pmatrix} \right) \tag{3.93}$$

These restrictive assumptions make it possible to point towards the differences between the Gaussian case and the case of a true mixture distribution more clearly, they enable us to better work out the nature of the model, and lead to simpler formulas as opposed to the general case. Moreover, it is this particular structure for model matrices that will be employed for the empirical application in chapter 9. Computation of the moments is delegated to the appendix.[55]

We first provide moments of the factors. Since yields are affine functions of them, moments of yields depend on the factor moments in a simple way.

Obviously, the process $\{X_t\}$ has expectation zero, i.e.

$$E(X_{it}) = 0 \tag{3.94}$$

for each element of X_t.

Second moments are given as

$$E(X_{it}^2) = \frac{E(u_{it}^2)}{1-\kappa_i^2}, \tag{3.95}$$

and

$$E(X_{it}X_{jt}) = 0 \text{ for } i \neq j. \tag{3.96}$$

That is, under our assumptions, two different factors are uncorrelated in the Gaussian case and in the mixture case.

[55] Moments for any linear multifactor model, that is not nested within these special assumptions, can be computed along similar lines. The derivations may be more cumbersome, but will make use of the same types of techniques and arguments as used for the model under our simplifying assumptions.

We have assumed that all μ_b are zero, so the distribution of u_t is symmetric around zero. As a consequence, all third moments vanish,

$$E(X_{it}X_{jt}X_{kt}) = 0. \tag{3.97}$$

Allowing the μ_b in the distribution of the factor innovations to be different from zero can induce asymmetry in the distribution of u_t. This would in turn cause nonzero third moments of the factors and skewness in bond yields. Moreover, it would also affect second and fourth moments. If $\mu_b \neq 0$, two innovations u_{it} and u_{jt} need not be uncorrelated anymore, which in turn implies that the corresponding factors are not uncorrelated anymore. We do not analyze the case $\mu_b \neq 0$ in more depth in this section and focus on fourth moments, i.e. on the possibility of excess kurtosis, instead.

For fourth moments, we have that

$$E(X_{it}^3 X_{jt}) = 0, \quad i \neq j, \tag{3.98}$$

$$E(X_{it}^2 X_{jt} X_{kt}) = 0, \quad i,\ j,\ k \text{ are all different}, \tag{3.99}$$

$$E(X_{it}X_{jt}X_{kt}X_{lt}) = 0, \quad i,\ j,\ k,\ l \text{ are all different}. \tag{3.100}$$

Positive expressions only result from index combinations of the form $E(X_{it}^4)$ and $E(X_{it}^2 X_{jt}^2)$. We have

$$E(X_{it}^4) = \frac{6\kappa_i^2 E(X_{it}^2)E(u_{it}^2) + E(u_{it}^4)}{1-\kappa_i^4}, \tag{3.101}$$

and

$$E(X_{it}^2 X_{jt}^2) = \frac{\kappa_i^2 E(X_{it}^2)E(u_{jt}^2) + \kappa_j^2 E(X_{jt}^2)E(u_{it}^2) + E(u_{it}^2 u_{jt}^2)}{1-\kappa_i^2\kappa_j^2}, \quad i \neq j. \tag{3.102}$$

This is the point where the moments of a VAR(1) with Gaussian innovations and the moments of a VAR(1) with innovations from a Gaussian mixture (that has the same second moments) differ. As shown[56] in section 3.2.2, the fourth moment of the ith innovation, $E(u_{it}^4)$, from a Gaussian mixture is greater than that of a simple normal.[57] Thus, $E(X_{it}^4)$ for the mixture case exceeds its counterpart from the simple normal.

In the Gaussian case, the term $E(u_{it}^2 u_{jt}^2)$ equals $E(u_{it}^2)E(u_{jt}^2)$. For factors, this implies that $E(X_{it}^2 X_{jt}^2) = E(X_{it}^2)E(X_{jt}^2)$. However, this is not true for the mixture variant, where $E(u_{it}^2 u_{jt}^2) \neq E(u_{it}^2)E(u_{jt}^2)$ unless either the v_{ib}^2,

[56] There we have shown that the excess kurtosis for a mixture distribution is positive. This in turn implies that the fourth moment of the mixture is bigger than that of a simple normal, when they share the same variance.

[57] Unless $v_{ib} = v_i$, for all b, then the fourth moments coincide.

the v_{jb}^2 or both are all the same. That is, two factors X_{it} and X_{jt} are contemporaneously uncorrelated but not independent in the mixture case. This is a difference compared to the pure Gaussian situation.

Turning now to the computation of the moments of yields, we recall that for the mixture d-factor model, yields are given as[58]

$$y_t^n = \frac{A_n}{n} + \sum_{i=1}^{d} \frac{B_{in}}{n} X_{it}, \tag{3.103}$$

where B_{in} is the ith component of the vector B_n.

Since all factors are mean zero processes, the expectation of the n-period yield becomes

$$E(y_t^n) = \frac{A_n}{n}. \tag{3.104}$$

For the following we denote deviations from the mean yields by y_t^{n*}, i.e.

$$y_t^{n*} = y_t^n - E(y_t^n) = \sum_{i=1}^{d} \frac{B_{in}}{n} X_{it}.$$

The lth moment of the n-period yield around its mean is

$$E[(y_t^{n*})^l] = E\left[\left(\sum_{i=1}^{d} \frac{B_{in}}{n} X_{it}\right)^l\right] \tag{3.105}$$

Making use of the results that we have obtained for the moments of factors, we have for the variance

$$E[(y_t^{n*})^2] = \sum_{i=1}^{d} \left(\frac{B_{in}}{n}\right)^2 E(X_{it}^2), \tag{3.106}$$

where all cross products vanish due to (3.96). The covariance between the n-period yield and the m-period yield is

$$E[(y_t^{n*})(y_t^{m*})] = \sum_{i=1}^{d} \frac{B_{in}}{n} \frac{B_{im}}{m} E(X_{it}^2), \tag{3.107}$$

thus the contemporaneous correlation between two different yields is given by

$$Corr(y_t^n, y_t^m) = \frac{\sum_{i=1}^{d} B_{in} B_{im} E(X_{it}^2)}{\left(\sqrt{\sum_{i=1}^{d} B_{in}^2 E(X_{it}^2)}\right) \cdot \left(\sqrt{\sum_{i=1}^{d} B_{im}^2 E(X_{it}^2)}\right)}. \tag{3.108}$$

[58] See (3.79) above.

This is a difference to the one-factor models. There, all yields are perfectly correlated, which is a contradiction to the stylized facts. Here, in general, the correlation is different from one.

Obviously, third moments of yields vanish,

$$E[(y_t^{*n})^3] = 0 \tag{3.109}$$

i.e. there can be no skewness in bond yields for the case $\mu_b = 0$

For fourth moments, we have the expression

$$E[(y_t^{*n})^4] = \sum_{i=1}^{d} \sum_{j=1}^{d} \left(\frac{B_{in}}{n}\right)^2 \left(\frac{B_{jn}}{n}\right)^2 E(X_{it}^2 X_{jt}^2). \tag{3.110}$$

from which it is possible to derive the formula for excess kurtosis of yields. In order to do so, we first give a representation of fourth factor moments that decomposes them into a sum of two components: one that holds for the Gaussian case and another that captures the difference between the Gaussian and the mixture case.

Consider a Gaussian multifactor model and a mixture model that share the same second moments. Fourth moments of the corresponding factor processes differ from each other due to different fourth moments of the innovations u_t. Let

$$d_{ij} := E(u_{it}^2 u_{jt}^2) - E(u_{it}^2 u_{jt}^2)_{Gauss} \tag{3.111}$$

denote the difference between the fourth moments of the factor innovation for the mixture model and the Gaussian model as its special case. Then we can write

$$E(X_{it}^2 X_{jt}^2) = E(X_{it}^2 X_{jt}^2)_{Gauss} + \frac{1}{1 - \kappa_i^2 \kappa_j^2} d_{ij}.$$

With

$$w_{ij} := \frac{1}{1 - \kappa_i^2 \kappa_j^2}$$

it follows for the fourth moment of yields, that

$$\begin{aligned} E[(y_t^{n*})^4] &= \sum_{i=1}^{d} \sum_{j=1}^{d} \left(\frac{B_{in}}{n}\right)^2 \left(\frac{B_{jn}}{n}\right)^2 \left(E(X_{it}^2 X_{jt}^2)_{Gauss} + w_{ij} \cdot d_{ij}\right) \\ &= E[(y_t^{n*})^4]_{Gauss} + \sum_{i=1}^{d} \sum_{j=1}^{d} \left(\frac{B_{in}}{n}\right)^2 \left(\frac{B_{jn}}{n}\right)^2 w_{ij} \cdot d_{ij}. \end{aligned}$$

For the kurtosis of the mixture model we obtain

$$\begin{aligned} kurt(y_t^n) &= \frac{E[(y_t^{n*})^4]_{Mixt}}{(E[(y_t^{n*})^2])^2} - 3 \\ &= \frac{E[(y_t^{n*})^4]_{Gauss} + \sum_{i=1}^{d} \sum_{j=1}^{d} \left(\frac{B_{in}}{n}\right)^2 \left(\frac{B_{jn}}{n}\right)^2 w_{ij} \cdot d_{ij}}{(E[(y_t^{n*})^2])^2} - 3 \end{aligned}$$

Since the excess kurtosis of yields in the Gaussian model is zero, we have

$$kurt(y_t^n) = \frac{\sum_{i=1}^{d}\sum_{j=1}^{d}\left(\frac{B_{in}}{n}\right)^2\left(\frac{B_{jn}}{n}\right)^2 w_{ij}\cdot d_{ij}}{(E[(y_t^{n*})^2])^2}. \tag{3.112}$$

We observe, that the kurtosis of bond yields depends on the parameters κ_i, the parameters of the distribution of u_t, and time to maturity n. Unlike with the one-factor models, maturity-dependent terms do not cancel, so the kurtosis changes with time to maturity, a feature that is observed in the data.

The computation of moments of differenced yields and autocorrelations is delegated to the appendix. Unlike the one-factor models, the multifactor models exhibit an autocorrelation that varies with time to maturity.

In this chapter we have introduced the class of affine multifactor Gaussian mixture (AMGM) models, a canonical representation has been introduced, and we have derived some of the properties of these models. It turns out that the multifactor model has a greater flexibility to capture features that are observed with empirical data. In chapter 9 we will confront selected models from the AMGM class with data on US interest rates and see how they perform.

4

Continuous-Time Models of the Term Structure

We introduce term structure models in continuous time. The exposition will be less detailed than the one given for discrete-time models. We present the stochastic setting and outline the martingale approach to bond pricing: absence of arbitrage requires that there is a probability measure under which all discounted bond prices become martingales. The class of exponential-affine multifactor models and the approach of Heath Jarrow and Morton are discussed.[1] For both types of term structure models, specific examples are presented.

The martingale approach is by far the most common in continuous-time finance. We could have also employed such an approach for our discrete-time models in the previous chapter. Conversely, one can could also represent the continuous-time models employing a continuous-time pricing kernel process. Choosing one over the other, is in both cases a matter of convenience with respect to model building.

4.1 The Martingale Approach to Bond Pricing

Unlike in the last chapter the parameter set for the time parameter is the interval $[0, T^*]$, where $T^* < \infty$ is a time horizon such that only market activities before T^* are of interest. The stochastic setting that will be used in this section is a probability space $(\Omega, \mathcal{F}, \mathbb{P})$ and a filtration of sub σ-algebras $\mathbb{F} = \{\mathcal{F}_t : 0 \leq t \leq T^*\}$ with $\mathcal{F}_s \subseteq \mathcal{F}_t \subseteq \mathcal{F}$ for $s < t$. This filtration governs the evolution of information in the models discussed below. A stochastic process$\{X_t\} = \{X_t, t \in D \subset [0, T^*]\}$ is a family of random vectors. If not otherwise stated we assume that $D = [0, T^*]$. If X_t is $\mathcal{F}_t$-measurable for each t, the stochastic process is said to be adapted to $\mathbb{F}$.

[1] The presentation of the material is mostly based on [14], [17], [19], [43] and [87]. See also the recent overview by Piazzesi [92].

The source of randomness is a standard d-dimensional $\mathbb{P}$-Brownian motion $W = \{W_t\}$, with $W_t = (W_t^1, \ldots, W_t^d)'$ which is adapted to $\mathbb{F}$. The probability measure $\mathbb{P}$ reflects the subjective probability of market participants concerning the events in $\mathcal{F}$. As such, $\mathbb{P}$ is commonly referred to as the real-world measure or the physical measure.[2] By $E^{\mathbb{M}}$ we denote the expectation operator with respect to the probability measure $\mathbb{M}$. The superscript becomes necessary, since other probability measures besides $\mathbb{P}$ will be worked with below. Conditional expectations will be written in the form $E^{\mathbb{M}}(X_s|\mathcal{F}_t)$, which may be abbreviated as $E_t^{\mathbb{M}}(X_s)$.

The notion of arbitrage is similar to that in discrete time. Again, there are various definitions in the literature, most of them turn out to be equivalent. We summarize the definition given in [17], leaving all technical details aside. Consider a market with $N+1$ assets each characterized by its price process $S = \{S_t\}$ with $S_t = (S_{0t}, S_{1t}, \ldots, S_{Nt})'$. The first component is assumed to be a numeraire process, i.e. it satisfies $S_{0t} > 0$ for all t. The process $\{\tilde{S}_{it}\}$ with $\tilde{S}_{it} = S_{it}/S_{0t}$ is called the discounted price process of asset i. That is, at each t the price of asset i is expressed in terms of the numeraire asset. A trading strategy $H = \{H_t\}$ is a vector-valued adapted process such that $H_t = (H_{0t}, H_{1t}, \ldots, H_{Nt})' \in \mathbb{R}^{N+1}$, represents the quantities of assets held in a portfolio at time t. At time t, the value of the portfolio is equal to $V_t(H) = H_t'S_t$, accordingly, the corresponding stochastic process $V(H) = \{V_t(H)\}$ is referred to as the value process. A trading strategy H is called self-financing if the value process satisfies

$$V_t(H) = V_0(H) + \sum_{i=0}^{N} \int_0^t H_{it}\, dS_{it}.$$

A trading strategy is called an arbitrage strategy (or arbitrage for short) if it is self-financing and if its value process satisfies

$$V_0(H) = 0, \quad \mathbb{P}(V_T \geq 0) = 1, \quad \text{and } \mathbb{P}(V_T(H) > 0) > 0.$$

Concerning the bond market, following [18], it is assumed that there is a frictionless market for any T-bond with $T \in (0, T^*]$. Using the core argument of financial arbitrage theory, the bond market is free of arbitrage if there is a probability measure $\mathbb{Q}$ equivalent to $\mathbb{P}$ such that all bond price processes - discounted with respect to some numeraire B_t - are martingales under $\mathbb{Q}$.

Note that, strictly speaking, the notion of arbitrage introduced above is not applicable to a bond market that is characterized by a continuum of maturities and thus contains an infinite number of assets. However, the existence of an equivalent martingale measure as described above implies that *all* bond price processes are martingales. That is, any finite family of bond price processes satisfies the condition of no arbitrage.

[2] See, e.g. [87].

From the martingale condition, one obtains immediately that an arbitrary T-bond has to satisfy:

$$E^{\mathbb{Q}}\left(\frac{P(T,T)}{B_T}\bigg|\mathcal{F}_t\right)=\frac{P(t,T)}{B_t}.$$

Employing that $P(T,T)=1$, this yields

$$P(t,T)=B_t\cdot E^{\mathbb{Q}}\left(\frac{1}{B_T}\bigg|\mathcal{F}_t\right). \tag{4.1}$$

For a specific choice of the numeraire we assume that the short rate r_t is an adapted process that satisfies

$$\int_0^{T^*} r_t^2 dt<\infty$$

and define the money account process $\{B_t\}$ by

$$B_t=\exp\left(\int_0^t r_s ds\right). \tag{4.2}$$

B_t is interpreted as the time t value of a unit of account invested at time 0 and continously compounded at rate r. Using (4.2) as numeraire, the basic bond pricing formula in (4.1) becomes

$$P(t,T)=E^{\mathbb{Q}}\left(\exp(-\int_t^T r_s ds)\bigg|\mathcal{F}_t\right). \tag{4.3}$$

That is, the price of the T-bond at time t is the expected discounted payoff (equal to one), where the discounting uses future short-term interest rates and the expectation is with respect to the martingale measure $\mathbb{Q}$.

We will now briefly turn to the following two important questions. First: how is the martingale measure $\mathbb{Q}$ constructed from the physical measure $\mathbb{P}$? Second, given standard $\mathbb{P}$-Brownian motion W, how can one construct a process $\tilde{W}$ that is standard Brownian motion under $\mathbb{Q}$?[3]

Assume there exists a d-dimensional process $\lambda=\{\lambda_t\}$ that satisfies

$$E\left[\exp\left(\frac{1}{2}\int_0^t|\lambda_s|^2 ds\right)\right]<\infty,$$

which is known as Noikov's condition. From λ a process $L=\{L_t\}$ is constructed as

$$L_t=\exp\left(\int_0^t -\lambda_s dW_s-\frac{1}{2}\int_0^t|\lambda_s|^2 ds\right). \tag{4.4}$$

[3] For the following see [43] and the references given therein.

It can be shown that L is a martingale. Now a measure $\mathbb{Q}$ equivalent to $\mathbb{P}$ can be constructed using L_T as Radon-Nikodym derivative, i.e.

$$\mathbb{Q}(F) = \int_F L_T \, d\mathbb{P}, \quad \text{for } F \in \mathbb{F}. \tag{4.5}$$

In light of the latter we will sometimes write $\mathbb{Q}(\lambda)$ when it should be emphasized based on which process λ the measure $\mathbb{Q}$ is constructed.

Finally, Girsanov's theorem provides the following relation. Let W be a standard $\mathbb{P}$-Brownian motion and let λ and $\mathbb{Q}(\lambda)$ be as given above. Define the process $\tilde{W} = \{\tilde{W}\}$ by

$$\tilde{W}_t = W_t + \int_0^t \lambda_s ds, \tag{4.6}$$

or in differential notation

$$d\tilde{W}_t = dW_t + \lambda_t dt. \tag{4.7}$$

Then the process $\tilde{W}$ is standard Brownian motion under the measure $\mathbb{Q}(\lambda)$.

4.1.1 One-Factor Models of the Short Rate

Going from the general to the specific, suppose now that the dynamics of the short rate process $r = \{r_t\}$ is governed by a stochastic differential equation (SDE) of the form[4]

$$dr_t = \mu(r_t, t, \psi) \, dt + \sigma(r_t, t, \psi) \, dW_t, \tag{4.8}$$

where the functions $\mu(\cdot)$ and $\sigma(\cdot)$ satisfy a set of regularity conditions. $\mu(\cdot)$ and $\sigma(\cdot)$ are called the drift and diffusion of the SDE, respectively. ψ is a vector of parameters, and W is one-dimensional $\mathbb{P}$-Brownian motion. Specific functional forms for $\mu(\cdot)$ and $\sigma(\cdot)$ define particular one-factor short rate models, examples of which are given in section 4.2.2.

With a particular model for the short rate at hand and having no further exogenously given security price process, it is in principle now a straightforward matter to find an arbitrage-free family of bond price processes: one can solve the SDE (4.8) and then evaluate the expectation of the integral in (4.3). However, such explicit computation is only possible for specific functional forms of $\mu(\cdot)$ and $\sigma(\cdot)$.

For an alternative approach, we first write the solution for the bond price as an explicit function of the short rate process r and t,

$$P(t, T) = F^T(r, t).$$

[4] Note that the SDE is a short hand notation for the stochastic integral expression $r_t = r_0 + \int_0^t \mu(r_s, s) ds + \int_0^s \sigma(r_s, s) dW_s$.

We will refer to F^T as the pricing function. If the solution for the price of a T-bond solves (4.3) under (4.8) for every t, it will also solve the following deterministic partial differential equation (PDE)

$$\frac{\partial F^T(r,t)}{\partial t} + (\mu - \sigma\lambda_t)\frac{\partial F^T(r,t)}{\partial r_t} + \frac{1}{2}\sigma^2\frac{\partial^2 F^T(r,t)}{\partial r_t^2} - r_t F^T(r,t) = 0, \quad (4.9)$$

where the arguments of μ and σ have been suppressed. This equivalence is provided by a theorem known as the Feynman-Kac connection.[5] Making use of the latter relation, finding bond prices means solving the PDE (4.9).

For the following we make use of the following short hand notation: we denote partial derivatives by subscripts and omit the arguments of F^T. If this short hand notation for partial derivatives is used, we will drop time subscripts in order to avoid confusion. That is, (4.9) is denoted in short by

$$F_t^T + (\mu - \sigma\lambda)F_r^T + \frac{1}{2}\sigma^2 F_{rr}^T - rF^T = 0. \quad (4.10)$$

Next, we will examine the role of λ - the process connecting the measures $\mathbb{P}$ and $\mathbb{Q}$ via (4.5) and (4.4) - a little closer.

The solution for the price process of the T-bond is a function of r_t and time t. Since the SDE for r_t is given by (4.8), we can find the SDE for the pricing function of the T-bond under the measure $\mathbb{P}$ using Ito's lemma. We obtain

$$dF^T = (F_t^T + \mu F_r^T + \frac{1}{2}\sigma^2 F_{rr}^T)dt + \sigma F_r^T dW. \quad (4.11)$$

Note that the partial derivatives in the Ito relation above also appear in the PDE (4.10), that defines the solutions of the bond price processes. It implies that the drift term in (4.11) can be written as $rF^T + \sigma\lambda F_r^T$. Inserting this into (4.11) and dividing by F^T we obtain an SDE describing the evolution of the instantaneous return of the T-bond,

$$\frac{dF^T}{F^T} = \left(r + \sigma\lambda\frac{F_r^T}{F^T}\right)dt + \sigma\frac{F_r^T}{F^T}dW. \quad (4.12)$$

Hence, the mean m and standard deviation s of the return on the T-bond are given as

$$m = r + \sigma\lambda\frac{F_r^T}{F^T} \quad \text{and} \quad s = \sigma\frac{F_r^T}{F^T}$$

respectively. When we compute the Sharpe ratio, i.e. the expected *excess* return of the T-bond over the risk-free short rate per unit of standard deviation, this is exactly λ, i.e. $(m - r)/s = \lambda$. Because of this property, λ is referred to as the market price of risk.

[5] See, e.g., [74].

4.1.2 Comments on the Market Price of Risk

Concerning the market price of risk, a few comments are in order. First, we observe that λ is not dependent on maturity T. This implies that in an arbitrage-free bond market all bonds must have the same market price of risk. Second, recall that λ is a stochastic process. That is, although being equal for all bonds at a given time t ('in the cross section') the market price of risk may well change over time. Third, λ links $\mathbb{P}$-Brownian motion and $\mathbb{Q}$-Brownian motion via (4.7). Inserting this into (4.12) yields an SDE for the bond price under the martingale measure $\mathbb{Q}$:

$$\frac{dF^T}{F^T} = rdt + \sigma \frac{F_r^T}{F^T} d\tilde{W}. \tag{4.13}$$

Thus, under $\mathbb{Q}$ all bonds have expected return equal to the short rate. Fourth, consider what happens when λ is zero. In this case, the measure $\mathbb{P}$ coincides with the martingale measure $\mathbb{Q}$ as can be seen from (4.5)-(4.4). Accordingly, bond prices can now be computed as the expectation of the discounted payoff (1 unit of account) under the real world measure $\mathbb{P}$ in (4.3). If this case prevails, it will be said that the *pure expectations hypothesis* of the term structure holds.[6]

Finally, there remains the natural question of how λ is determined or - equivalently - as Björk [19], p. 254, puts it: "Who chooses the martingale measure" $\mathbb{Q}(\lambda)$? Recall that bond price processes depend on the functions $\mu(\cdot)$ and $\sigma(\cdot)$ of the short rate process (4.8) but also on the process λ. The latter however, is not given by our pre-specified interest rate process. Thus, bond prices are only unique up to a specific choice of λ or equivalently $\mathbb{Q}(\lambda)$. The reason for $\mathbb{Q}$ not being unique is that we are considering an incomplete market: there is one source of randomness, our $\mathbb{P}$-Brownian motion W, and zero tradable assets. In fact, r_t is not the price of a tradable asset and therefore one cannot construct any portfolio; all one can do is invest money at interest rate r. Consequently, one cannot tie down the price of a particular T-bond by postulating that it must be equal to the value of a portfolio that replicates its payoff: such a portfolio does simply not exist in this framework. Given (4.8), choosing a particular process λ^1 and pricing bonds according to (4.3) leads to a family of bond prices that satisfies the "internal consistency"[7] relation of no arbitrage. Choosing another process λ^2 also generates an arbitrage-free family, but the particular solutions for the bond price processes are different from those belonging to λ^1.

It should be emphasized that the situation would be different when - in addition to the short rate process specified above - the process of a particular S-bond would also be given exogenously. This would create a complete market where the market price of risk process is uniquely determined.[8]

---

[6] See [87].
[7] [19, p. 245].
[8] See [19].

Using the econometric techniques that will be described in the sections below, it is possible to estimate the market price of risk process from a panel of bond prices. The corresponding martingale measure $\mathbb{Q}(\lambda)$ can then be interpreted as the one that is actually chosen by the market.[9]

4.1.3 Multifactor Models of the Short Rate

In the one-factor model described above, the Brownian motion driving the short rate process has been the only source of randomness. In turn, the short rate r_t can be regarded as a sufficient statistic for all bond prices. Multifactor models extend this framework by letting the short rate be driven by more than one factor. The general set up is as follows. A d-dimensional factor process $X = \{X_t\}$ is defined by the SDE

$$dX_t = \mu(X_t, t; \psi)\, dt + \sigma(X_t, t; \psi)\, dW_t, \tag{4.14}$$

where W is d-dimensional $\mathbb{P}$-Brownian motion that may be contemporaneously correlated; $\mu(\cdot)$ and $\sigma(\cdot)$ are functions that satisfy conditions such that the solution of (4.14) exists. Note that $\mu(\cdot)$ is now a $d \times 1$ vector and $\sigma(\cdot)$ is a $d \times d$ matrix. The short rate at t depends on X_t and t via a function f_r,

$$r_t = f_r(X_t, t; \psi). \tag{4.15}$$

The one-factor model discussed above is therefore the special case in which the short rate itself coincides with the only factor.

Bond prices are again obtained by the valuation formula (4.3). The quantity λ linking the measures $\mathbb{P}$ and $\mathbb{Q}$ via the Radon-Nikodym derivative in (4.4) is now vector-valued.

The solution for bond prices is now a function of the vector valued factor process X, i.e. $P(t,T) = F^T(X,t)$. The PDE for the pricing function is analogous to (4.10). For the multifactor case we have

$$\frac{1}{2} tr(\sigma\sigma' F^T_{XX}) + (\mu - \sigma\lambda)' F^T_X + F^T_t - r\,F^T = 0, \quad F^T(X,T) = 1. \tag{4.16}$$

Similar to the univariate case in the previous section, the variable λ can be interpreted as as a market price of risk vector. In analogy to the explanation in the one-dimensional case, the SDE of the bond price under $\mathbb{P}$ is of the form

$$\frac{dF^T}{F^T} = m dt + s' dW = m dt + \sum_{i=1}^{d} s_i dW_i.$$

For any T-bond one obtains the result that

---

[9] "Question: Who chooses the martingale measure? Answer: The market!", [19], p. 254.

$$m - r = \sum_{i=1}^{d} \lambda_i s_i. \tag{4.17}$$

Thus, the ith component of λ indicates how much one unit of the ith volatility component contributes to the risk premium of any T-bond over the short-term interest rate.

4.1.4 Martingale Modeling

Up to now we have specified the dynamics of the short rate under the real world measure $\mathbb{P}$. We then linked it to the martingale measure $\mathbb{Q}$ using the market price of risk process λ. An alternative strategy consists of directly modeling the short rate under $\mathbb{Q}$. For the factor process in the multifactor model we have

$$dX_t = \tilde{\mu}(X_t, t; \psi)\, dt + \sigma(X_t, t; \psi)\, d\tilde{W}_t, \tag{4.18}$$

where $\tilde{W}$ is d-dimensional $\mathbb{Q}$-Brownian motion. $\mathbb{P}$ and any $\mathbb{Q}(\lambda)$-Brownian motion are related by (4.7). Accordingly, the SDEs (4.14) and (4.18) have the same diffusion σ but different mean processes μ and $\tilde{\mu}$ which are related by

$$\tilde{\mu} = \mu - \sigma\lambda. \tag{4.19}$$

Using this relation, the fundamental PDE for the bond price in (4.16) becomes

$$\frac{1}{2} trace(\sigma\sigma' F^T_{XX}) + \tilde{\mu}' F^T_X + F^T_t - rP = 0, \quad F^T(X,T) = 1 \tag{4.20}$$

4.2 The Exponential-Affine Class

In this subsection we discuss the exponential-affine class of continuous-time term structure models. For models of this class, yields will be affine functions of the factors. The SDE of the factor process is characterized by a linear drift term, volatility may be level-dependent. The linear (purely) Gaussian multifactor models that have been discussed in the previous chapter are the discrete-time analogy to these continuous-time models (for the case in which volatility is not level-dependent).[10]

4.2.1 Model Structure

Under the condition of no arbitrage, bond prices must be equal to the conditional expectation in (4.3) or equivalently they must solve the PDE (4.20). We will now discuss a category of models for which the solution

[10] This chapter will not cover a continuous-time analogy to the discrete-time *mixture* models.

$P(t,T) = F^T(X,t)$ for the price process of a T-bond can be written explicitly as a simple function of the factors, namely as

$$P(t,T) = \exp\Big(A(n) + B(n)'X_t\Big). \tag{4.21}$$

$A(n)$ and $B(n)$ are functions of time to maturity $n = T - t$ and the model parameters. $A(n)$ is a scalar, whereas $B(n)$ has dimension $d \times 1$. Accordingly, using (2.2), yields are affine functions of the factors,

$$y(t,T) = \frac{1}{n}\Big(A(n) + B(n)'X_t\Big). \tag{4.22}$$

As shown by [45][11] - under certain technical conditions - bond prices are of the form (4.21) if and only if the short rate model under $\mathbb{Q}$, (4.15), (4.18), has the following structure:[12]

$$dX_t = \tilde{\mathcal{K}}(\tilde{\theta} - X_t)dt + \Sigma\sqrt{S_t}dW_t, \tag{4.23}$$

where S_t is a diagonal matrix whose ith element is given by

$$S_{ti} = \alpha_i + \beta_i'X_t. \tag{4.24}$$

That is, the drift and volatility are affine in the factors. The d $\mathbb{Q}$-Brownian motions are independent. The function f_r in (4.15) that links the short rate to the factors has to be affine as well,

$$f_r(X_t,t) = \delta + \gamma'X_t. \tag{4.25}$$

In order to see that under these conditions bond prices are of the form (4.21) one proceeds as follows. Take a short rate model satisfying (4.23) - (4.25). Guess that the solution to the PDE (4.20) is of the form (4.21).[13] Compute the required partial derivatives and insert them into (4.20). It turns out that (4.21) is in fact a solution to (4.20), where the coefficients A and B have to satisfy certain conditions that are given as a system of ordinary differential equations:[14]

$$\frac{dA(n)}{dn} = -\delta - \tilde{\theta}'\tilde{\mathcal{K}}'B(n) + \frac{1}{2}\sum_{i-1}^{d}[(\Sigma'B(n))_i]^2\alpha_i \tag{4.26}$$

$$\frac{dB(n)}{dn} = \gamma - \tilde{\mathcal{K}}'B(n) + \frac{1}{2}\sum_{i=1}^{d}[(\Sigma'B(n))_i]^2\beta_i \tag{4.27}$$

---

11 See also [44].

12 More interpretation of this model structure will be given below within the description of specific models from this class.

13 That is, the solution function F^T is given by the right hand side of (4.21).

14 $(\Sigma'B(n))_i$ denotes the ith element of the $d \times 1$ vector $\Sigma'B(n)$.

The boundary condition $P(T,T) = 1$ imposes the restriction that

$$A(0) = 0, \quad B(0) = 0. \tag{4.28}$$

The conditions for the equivalence between bond prices being exponential-affine in the factors and the short rate model being of the form (4.23) - (4.25) were formulated within the martingale approach: the constraints on drift and diffusion in (4.23) and (4.24) were formulated for an SDE of the factors under a martingale measure $\mathbb{Q}$. Are there similar conditions that must hold for the factor process under the real world measure $\mathbb{P}$? It should be recalled that the change of measure only affects the drift while leaving the diffusion unchanged. Thus, the conditions for the diffusion matrix of the factor process are the same under $\mathbb{P}$. The drift μ under $\mathbb{P}$ and the drift $\tilde{\mu}$ under $\mathbb{Q}$ are linked by the market price of risk λ in (4.19). Hence, the condition that the 'risk neutral drift' $\tilde{\mu}$ has to be affine in the factors imposes a *joint* restriction on the real world drift μ and λ.

If the factor process under $\mathbb{P}$ is already of the affine form, that is

$$dX_t = \mathcal{K}(\theta - X_t) + \Sigma\sqrt{S_t}dW_t, \tag{4.29}$$

with S_t as in (4.24), then λ must be chosen in a way that the drift remains affine under $\mathbb{Q}$. One can show that the process λ has to satisfy

$$\lambda_t = \sqrt{S}\,h, \tag{4.30}$$

for some $d \times 1$ vector of constants h. Then the factor process under $\mathbb{Q}$ has the required affine form and the risk-neutral parameters in (4.23) can be computed from the parameters in (4.29) and (4.30) as[15]

$$\tilde{\mathcal{K}} = (\mathcal{K} - \Sigma\Phi), \tag{4.31}$$

$$\tilde{\theta} = \tilde{\mathcal{K}}^{-1}(\mathcal{K}\theta - \Sigma\Psi) \tag{4.32}$$

where $\Phi = (h_1\beta_1', \dots, h_d\beta_d')'$ and $\Psi = (h_1\alpha_1, \dots, h_d\alpha_d)'$.

4.2.2 Specific Models

We will briefly summarize some models from the exponential-affine class.[16] If not otherwise stated, the dynamics of the factors are given under the real world measure $\mathbb{P}$. Further explanations are given in the original papers or in textbooks such as [87] and [65].

Two of the most prominent one-factor models are those by Vasiček [113] and Cox, Ingersoll and Ross [34] (CIR). In the Vasiček model,

[15] See [35].

[16] All of the models presented here have been estimated in the literature using the state space framework. Chapter 8 below contains references to the corresponding articles. Here, we focus on the structure of the theoretical models only.

$$dr_t = \kappa(\theta - r_t)dt + \sigma dW_t \tag{4.33}$$

and the CIR-model

$$dr_t = \kappa(\theta - r_t)dt + \sigma\sqrt{r_t}dW_t \tag{4.34}$$

the single factor coincides with the short-term interest rate r_t.

The parameter θ has the interpretation of the long run mean or equilibrium level of the short rate and κ governs how fast the interest rate is pulled back to equilibrium. Values of κ close to zero correspond to a short rate process which is close to being nonstationary, whereas higher values correspond to strong mean reversion.[17]

The assumption of constant volatility in the Vasiček model is dropped in the CIR model where the short rate's volatility depends on its level: if the current short rate r_t is high, the instantaneous volatility will also be high. Interest rates in the Vasiček world may well become negative whereas the CIR specification guarantees positivity.[18]

The multifactor version of the CIR model is given by

$$dX_{jt} = \kappa(\theta_j - X_{tj})dt + \sigma_j\sqrt{X_{jt}}dW_{jt}, \quad j = 1, \dots, d. \tag{4.35}$$

where the d factors are assumed to be independent.

The two-factor model by [15]

$$dr_t = \kappa_1(\theta_t - r_t)d_t + \sigma_1 dW_{1t} \tag{4.36}$$
$$d\theta_t = \kappa_2(\bar{\theta} - \theta_t)dt + \sigma_2 dW_{2t} \tag{4.37}$$

is also known as the double-decay model or as a model with stochastic central tendency. This is due to the fact that the equilibrium level of the short rate in (4.33) now becomes a random variable itself.

The specification

$$dr_t = \kappa_1(\theta_1 - r_t)d_t + \sqrt{v_t}dW_{1t} \tag{4.38}$$
$$dv_t = \kappa_2(\theta_2 - v_t)dt + \phi\sqrt{v_t}dW_{2t}. \tag{4.39}$$

can be interpreted as an extension of the CIR-model in which the evolution of the stochastic diffusion term $\sqrt{v_t}$ is itself governed by a square root process.

The three-factor model by [30],

$$dr_t = \kappa_1(\theta_t - r_t)d_t + \sqrt{v_t}dW_{1t} \tag{4.40}$$
$$d\theta_t = \kappa_2(\bar{\theta} - \theta_t)dt + \xi\sqrt{\theta_t}dW_{2t} \tag{4.41}$$
$$dv_t = \kappa_3(\bar{v} - v_t)dt + \mu\sqrt{v_t}dW_{3t}, \tag{4.42}$$

---

17 Heuristically, this can be reconciled with the notion of stationarity of a discrete-time AR(1) process. Write a discretization of (4.33) as $\Delta r_t = \kappa(\theta - r_{t-1}) + u_t$, or $r_t = \kappa\theta + (1 - \kappa)r_{t-1} + u_t$. Thus, nonstationarity prevails if $1 - \kappa = 1$ i.e. if $\kappa = 0$.

18 See [34].

can be regarded as a natural combination of the latter two-factor models. It exhibits a process for the stochastic volatility term as well as a stochastic central tendency for the short rate.

In all of the models discussed up to now, the short rate is given as an affine function of the factors. All models are nested within the exponential-affine class described above. Note that in all of the described models the first factor is identified as the short rate. For (4.25) this implies that $\delta = 0$ and $\gamma = (1, 0, \ldots, 0)'$. Similarly to our considerations for the discrete-time models one can argue that this parameterization can be achieved by starting from an arbitrary model and applying suitable invariant transformations.

4.3 The Heath-Jarrow-Morton Class

We will now present an approach to term structure modeling that takes the whole forward rate curve as starting point. The forward rate curve at some time t is the instantaneous forward rate $f(t,T)$ considered as a function of T. The evolution of the forward rate for any fixed $T < T^*$ is specified by the SDE

$$df(t,T) = \alpha(t,T)dt + \sigma(t,T)'dW_t \tag{4.43}$$

where W_t is d-dimensional $\mathbb{P}$-Brownian motion. It is further assumed that the initial forward curve at time $t = 0$ matches the observed forward rate curve $\bar{f}(0,T)$, i.e.

$$f(0,T) = \bar{f}(0,T) \tag{4.44}$$

for all T.

In the multifactor models of the previous section, the term structure of interest rates was driven by the d-dimensional process X. Note that in the framework considered here, the whole family of forward rate curves is exogenously given and that T can assume a continuum of values. Thus, this approach can be interpreted as an extension of the multifactor framework where the exogenously specified stochastic process is now infinite dimensional.[19]

Bond prices and forward rates are related by formula (2.7). Thus, given the continuum of forward rates at time t, the whole term structure at time t is specified. In order to rule out arbitrage opportunities, the processes in (4.43) must be restricted, i.e. one must impose conditions on $\alpha(t,T)$ and $\sigma(t,T)$. It can be shown that when the market is arbitrage-free there must be a d-dimensional process λ such that for all $t \leq T$ and all $T \geq 0$, the drift and diffusion terms in (4.43) are related as:[20]

$$\alpha(t,T) = \sigma(t,T)' \int_t^T \sigma(t,s)\, ds - \sigma(t,T)'\lambda_t. \tag{4.45}$$

---

[19] See [17].

[20] See Bingham and Kiesel [17] who also state the exact technical conditions that λ_t must satisfy.

These requirements for the absence of arbitrage have been developed by Heath, Jarrow and Morton [61] (HJM), and are therefore called 'HJM drift condtions'. When the forward rate process in (4.43) is already defined under a risk-neutral martingale measure $\mathbb{Q}$, then the process λ_t is identically zero, and the condition (4.45) becomes

$$\alpha(t,T) = \sigma(t,T)' \int_t^T \sigma(t,s)\,ds. \tag{4.46}$$

It should be emphasized that the HJM approach does not denote a specific model but rather a very flexible framework for the analysis of the term structure. Term structure modeling in this context can be summarized in a stylized fashion as follows:[21] Start with a specification of the volatility structure $\sigma(t,T)$. Construct the drift parameters $\alpha(t,T)$ via (4.46) which leads to an SDE for $f(t,T)$ under $\mathbb{Q}$. Solve the SDE using (4.44) as initial condition. With the solution $f(t,T)$ at hand, compute bond prices via (2.7).

In principle, models of the short rate as presented in the last section can be turned into HJM models and vice versa. [14] give a variety of examples for that. As an illustration, the one-factor model

$$dr_t = b(a - r_t)dt + c\,d\tilde{W}_t$$

for the short rate corresponds to a HJM model with volatility structure

$$\sigma(t,T) = ce^{-b(T-t)}$$

and initial forward rate curve given as

$$f(0,T) = a + e^{-bT}(r_0 - a) - \frac{c^2}{2b^2}\left(1 - e^{-bT}\right)^2.$$

This brings our exposition of discrete-time and continuous-time term structure models to an end. The models will show up again in chapter 8. There it will be shown that the statistical state space model can be used as a device for estimating the term structure models using a panel of observed interest rates.

[21] See [19].

5

State Space Models

This chapter introduces the statistical state space model. It is not intended to give an exhaustive treatment of the topic. We rather aim to provide a self-contained presentation of the statistical tools that will be employed for the estimation of term structure models.

We introduce the structure of state space models and present the statistical techniques for filtering, smoothing, prediction and parameter estimation. We leave aside the topics of diagnostics checking and model selection. These will be treated within the special context of chapter 8 below.

5.1 Structure of the Model

A state space model is a representation of the joint dynamic evolution of an observable random vector y_t and a generally unobservable state vector α_t.[1] It is a unifying framework that nests a wide range of dynamic models that are used in econometrics. The state space model contains a measurement equation and a transition equation. The transition equation governs the evolution of the state vector, the measurement equation specifies how the state interacts with the vector of observations. We first consider a quite general form of the state space model and then move on to more specializing assumptions concerning the functional forms and statistical distributions involved. All versions of state space models considered here will be in discrete time.[2]

Let α_t be an $r \times 1$ random vector. It will be called the state of the system or the state vector. Its evolution is governed by a dynamic process of the form

$$\alpha_t = T_t(\alpha_{t-1}) + \eta_t, \tag{5.1}$$

which is called the transition equation of the model. Given the realization of α_{t-1}, the conditional mean of α_t equals $T_t(\alpha_{t-1})$. The subscript t of the

[1] For expositions of the state space model, see [23], [46], [56], [57], [59] and [76].

[2] The framework that is described here may be extended to transition and/or measurement equations that are continuous-time diffusions, see, e.g., [101].

function $T_t(\cdot)$ denotes that this function may depend on time. The innovation vector η_t is a serially independent process with mean zero and finite variance-covariance matrix, which may also depend on time.

The measurement equation writes the $N \times 1$ vector y_t of observations as a (possibly time-dependent) function of the contemporaneous state α_t and an error term ϵ_t,

$$y_t = M_t(\alpha_t) + \epsilon_t. \tag{5.2}$$

The vector ϵ_t is also a serially independent process with mean zero and finite variance-covariance matrix.

For all state space models considered in this book, the random vectors η_t and ϵ_t are both assumed to be serially independent, and η_t and ϵ_s are independent for all s and t.

For both the transition equation and the measurement equation it is possible that exogenous or predetermined variables enter the functions $M_t(\cdot)$ and $T_t(\cdot)$ or the distributions of the state innovation and the measurement error. Concerning the applications of state space models in this book, the inclusion of additional explanatory variables will not play a role. Hence, they will not enter our model set-up in this chapter.

It is common for state space models to explicitly specify an initial condition, i.e. a distribution for the state vector at time $t = 0$,

$$\alpha_0 \sim D(\bar{a}_0, \bar{P}_0). \tag{5.3}$$

Finally, it is assumed that η_t and ϵ_t are independent of α_0 for all t.

Some preliminary comments about the functioning of the model are in order. Started by a draw from the distribution (5.3), the state vector evolves through time according to (5.1). The process has the Markov property. That is, the distribution of α_t at time t given the entire past realizations of the process, is equal to the distribution of α_t given α_{t-1} only. The state vector drives the evolution of the observation vector y_t. At each time t, the vector y_t is given as a sum of a systematic component and an error term: the systematic component is the transformation $M_t(\alpha_t)$ of the state vector. The mapping $M_t(\cdot)$ is from $\mathbb{R}^r$ to $\mathbb{R}^N$, and concerning the dimensions, all cases $r > N$, $r < N$, and $r = N$ are allowed. For $r < N$, e.g., the interpretation can be such that a low-dimensional state process drives a higher-dimensional observation process. The error ϵ_t drives a wedge between the systematic component $M_t(\alpha_t)$ and the observed vector y_t. The naming of ϵ_t as a measurement error stems from the use of the state space framework in the engineering or natural sciences. For instance, the state vector could contain the true coordinates of a moving object's location and information about its velocity. The observation vector may contain noisy measurements of the distance and the angle between the observer and the object.

In the context of the estimation of term structure models, treated in chapters 8 and 9 below, the observation vector will consist of (suitably transformed) functions of bond prices or interest rates for different times to maturity. This vector of observed financial variables will then be driven by a lower-dimensional vector of latent factors.

5.2 Filtering, Prediction, Smoothing, and Parameter Estimation

Associated with a state space model is the problem of estimating the unobservable state using a set of observations. Let $\mathcal{Y}_s$ denote a sequence of observations augmented by a constant vector y_0, i.e. $\mathcal{Y}'_s = (y'_0, y'_1, \ldots, y'_s)$. Assume further that the whole sample available for estimation is given by $\mathcal{Y}_T$. Without loss of generality, we assume that y_0 is a vector of ones, $y_0 = 1_N$. For fixed but arbitrary t we consider the problem of estimating α_t in terms of $\mathcal{Y}_s$. If $s = t$ the problem is called a filtering problem, $s < t$ defines a prediction problem, and the case $s > t$ is referred to as a smoothing problem. Besides predicting the unobservable state, one also considers the task of forecasting the observation vector y_t.

The mean squared error (MSE) will be used as the optimality criterion. Accordingly, the best estimators of the state vector are functions $\hat{\alpha}_t(\mathcal{Y}_s)$ that satisfy

$$E\left[(\alpha_t - \alpha_t(\mathcal{Y}_s))(\alpha_t - \alpha_t(\mathcal{Y}_s))'\right] \geq E\left[(\alpha_t - \hat{\alpha}_t(\mathcal{Y}_s))(\alpha_t - \hat{\alpha}_t(\mathcal{Y}_s))'\right] \tag{5.4}$$

for every function $\alpha_t(\mathcal{Y}_s)$ of $\mathcal{Y}_s$. The expectation is taken with respect to the joint density of $\mathcal{Y}_s$. The inequality sign denotes that the difference of the right hand side MSE-matrix and the left hand side MSE-matrix is negative semi-definite.

As is well known, the MSE-optimal estimator of α_t in terms of $\mathcal{Y}_s$ is given by the conditional expectation

$$\hat{\alpha}_t(\mathcal{Y}_s) = E(\alpha_t|\mathcal{Y}_s) = \int \alpha_t\, p(\alpha_t|\mathcal{Y}_s) d\alpha_t. \tag{5.5}$$

That is, for finding the optimal estimators $\hat{\alpha}_t(\mathcal{Y}_t)$ (filtered state), $\hat{\alpha}_t(\mathcal{Y}_{t-1})$ (predicted state) and $\hat{\alpha}_t(\mathcal{Y}_T)$ (smoothed state), one has to find the respective conditional densities $p(\alpha_t|\mathcal{Y}_t)$, $p(\alpha_t|\mathcal{Y}_{t-1})$ and $p(\alpha_t|\mathcal{Y}_T)$, and then compute the conditional expectation given by (5.5). Similarly, for obtaining $\hat{y}_t(\mathcal{Y}_{t-1})$, the optimal one-step predictor for the observation vector, one has to find the conditional density $p(y_t|\mathcal{Y}_{t-1})$ in order to compute $E(y_t|\mathcal{Y}_{t-1})$. We will refer to $p(\alpha_t|\mathcal{Y}_t)$, $p(\alpha_t|\mathcal{Y}_{t-1})$ and $p(y_t|\mathcal{Y}_{t-1})$ as filtering and prediction densities, respectively.

The following theorem states that the required conditional densities can be constructed in an iterative fashion.

Proposition 5.1. *Let $p(\alpha_{t-1}|\mathcal{Y}_{t-1})$ be given for some time $t-1$. Then*

$$p(\alpha_t|\mathcal{Y}_{t-1}) = \int p(\alpha_t|\alpha_{t-1})p(\alpha_{t-1}|\mathcal{Y}_{t-1})d\alpha_{t-1}, \tag{5.6}$$

$$p(y_t|\mathcal{Y}_{t-1}) = \int p(y_t|\alpha_t)p(\alpha_t|\mathcal{Y}_{t-1})d\alpha_t \tag{5.7}$$

$$p(\alpha_t|\mathcal{Y}_t) = \frac{p(y_t|\alpha_t)p(\alpha_t|\mathcal{Y}_{t-1})}{p(y_t|\mathcal{Y}_{t-1})}. \tag{5.8}$$

Proof. See [109]. □

Starting with $p(\alpha_0|\mathcal{Y}_0) = p(\alpha_0)$, equations (5.6) - (5.8) are iteratively applied. Note that the densities $p(\alpha_t|\alpha_{t-1})$ and $p(y_t|\alpha_t)$ that enter the equations above are implied by the transition equation (5.1) and the measurement equation (5.2) respectively. They will therefore be referred to as the transition density and the measurement density.

The conditional density $p(\alpha_t|\mathcal{Y}_T)$ is required to construct the (smoothing) estimator $E(\alpha_t|\mathcal{Y}_T)$. The sequence of densities $p(\alpha_t|\mathcal{Y}_T)$, $t = 1, \dots, T-1$, can be obtained by backward iteration. As shown in [109] the following relation holds:

$$p(\alpha_t|\mathcal{Y}_T) = p(\alpha_t|\mathcal{Y}_t) \int \frac{p(\alpha_{t+1}|\mathcal{Y}_T)p(\alpha_{t+1}|\alpha_t)}{p(\alpha_{t+1}|\mathcal{Y}_t)} d\alpha_{t+1} \tag{5.9}$$

Thus, given $p(\alpha_{t+1}|\mathcal{Y}_T)$ for some t, one can compute $p(\alpha_t|\mathcal{Y}_T)$. The conditional densities $p(\alpha_t|\mathcal{Y}_t)$ and $p(\alpha_{t+1}|\mathcal{Y}_t)$ have to be saved as results from the filtering iterations above. Taking $p(\alpha_T|\mathcal{Y}_T)$ as given from the filtering iterations, the first smoothing density that is computed by (5.9) is $p(\alpha_{T-1}|\mathcal{Y}_T)$. This is needed to compute $p(\alpha_{T-2}|\mathcal{Y}_T)$. In this fashion one proceeds until at last $p(\alpha_1|\mathcal{Y}_T)$ is computed.

Up to now it has been tacitly assumed that the state space model does not contain any unknown parameters. However, in almost all economic applications unknown parameters enter the measurement equation, the transition equation, and/or the distribution functions of the innovation η_t and the measurement error ϵ_t. Let the unknown parameters be collected in a vector ψ. We now show how ψ can be estimated by maximum likelihood.

Again, we take $\mathcal{Y}_T$ as the sequence of observations available for estimation. The joint density $p(y_1, \dots, y_T)$ can be written as the product of the conditional densities:

$$p(y_1, \dots, y_T) = \prod_{t=1}^{T} p(y_t|\mathcal{Y}_{t-1}). \tag{5.10}$$

The conditional densities $p(y_t|\mathcal{Y}_{t-1})$ are obtained within the iterative procedure in proposition 5.1. Thus, for a given ψ the iterations above can be used to compute the log-likelihood

$$l(\psi) = \sum_{t=1}^{T} \ln p(y_t|\mathcal{Y}_{t-1}; \psi). \tag{5.11}$$

Here, we have explicitly added ψ as an argument of $p(y_t|\mathcal{Y}_{t-1};\psi)$. Of course, the other conditional densities that show up in (5.6) - (5.8) will also be parameterized in ψ. Maximizing $l(\psi)$ with respect to ψ yields the maximum likelihood estimator $\hat{\psi}$.

The material presented up to here has dealt with the state space model using a quite general formulation. The algorithms described above are in principle the full answer to the problems of filtering, prediction, smoothing and parameter estimation. However, applying the results to a model of interest may induce some computational problems. Consider, for example, the task of computing the sequence of filtered states $\hat{\alpha_1}(\mathcal{Y}_1),\ldots,\hat{\alpha_T}(\mathcal{Y}_T)$. First, a large number of multiple integrals has to be computed within the algorithm (5.6) - (5.8). Second, with $p(\alpha_t|\mathcal{Y}_t)$ at hand for all t, one has to compute conditional expectations, i.e., the integrals $\int \alpha_t p(\alpha_t|\mathcal{Y}_s)d\alpha_t$. Computing these integrals for arbitrary functional forms $M_t(\cdot)$ and $T_t(\cdot)$ as well as for arbitrary distributions of η_t and ϵ_t will require heavy use of numerical methods.

For the most popular special case, the linear state space model with Gaussian state innovations and measurement errors, the filtering and prediction densities are all normal and can be computed using the Kalman filter which is discussed in the next subsection. Approaches to alleviate the computational problems for the general nonlinear and/or non-Gaussian case try to approximate the model itself or introduce simplifying approximations when computing the filtering and prediction densities.

Within the variety of methods, the extended Kalman filter is virtually the classic approach to handling nonlinearity in state space models.[3] With this approach, the functions $T_t(\cdot)$ and $M_t(\cdot)$ in (5.1) and (5.2) are linearized locally by a first degree Taylor approximation around the respective conditional means. A slightly modified version of the Kalman filter is applied to the resulting linearized system, yielding estimates of the conditional means.

Other approaches work with different types of linearization, use numerical integration for solving the integrals appearing in equations (5.6) - (5.8), or apply simulation-based techniques. For a survey of the latter see [110] and the references given therein.

The whole of the next chapter is devoted to linear state space models, for which the state innovation has a simple Gaussian or a Gaussian mixture distribution. This is particularly worthwhile since these kinds of state space models will turn out to be the natural statistical framework for the estimation of the AMGM term structure models introduced above. Some approaches to estimating nonlinear and non-Gaussian state space models will be presented within the context of the estimation of term structure models in chapter 8.

---

[3] A detailed account of the matter can be found in chapter 8 of [4]. See also [59].

5.3 Linear Gaussian Models

Against the background of the general set-up above, two restrictive assumptions are made: first, the functions $M_t(\cdot)$ and $T_t(\cdot)$ are affine. Second, the distributions of η_t, ϵ_t and α_0 are normal. Models satisfying these assumptions will be referred to as linear Gaussian state space models.

5.3.1 Model Structure

The transition equation is given by[4]

$$\alpha_t = T\alpha_{t-1} + c + \eta_t, \tag{5.12}$$

for the measurement equation we have

$$y_t = M\alpha_t + d + \epsilon_t. \tag{5.13}$$

The state innovation and measurement error are normally distributed,

$$\begin{pmatrix} \eta_t \\ \epsilon_t \end{pmatrix} \sim N\left(\begin{pmatrix} 0 \\ 0 \end{pmatrix}, \begin{pmatrix} Q & 0 \\ 0 & H \end{pmatrix} \right), \tag{5.14}$$

the initial condition becomes

$$\alpha_0 \sim N(\bar{a}_0, \bar{P}_0). \tag{5.15}$$

Finally,

$$E(\eta_t \alpha_0') = 0, \quad E(\epsilon_t \alpha_0') = 0, \text{ for all } t. \tag{5.16}$$

The quantities d, c, $\bar{a}_0$, M, T, H, Q and $\bar{P}_0$ are vectors and matrices of appropriate dimension. They will sometimes be referred to as the system matrices. Although it is not crucial, we assume that the system matrices are all constant over time. A linear Gaussian model with this property is referred to as time-homogeneous.

5.3.2 The Kalman Filter

For the model above, it follows that the transition density $p(\alpha_{t+1}|\alpha_t)$ and the measurement density $p(y_t|\alpha_t)$ are normal. It can be shown that this implies that also the prediction and filtering densities are normal:

$$\alpha_t|\mathcal{Y}_{t-1} \sim N(a_{t|t-1}, \Sigma_{t|t-1}) \tag{5.17}$$
$$\alpha_t|\mathcal{Y}_t \sim N(a_{t|t}, \Sigma_{t|t}) \tag{5.18}$$
$$y_t|\mathcal{Y}_{t-1} \sim N(\hat{y}_{t|t-1}, F_t). \tag{5.19}$$

[4] The matrix T in the transition equation is denoted by the same symbol as the number of observations. However, we think that it is always clear from the context which is referred to.

The normal densities are fully described by their first two moments. Thus, one has to find the sequences of conditional means, $a_{t|t-1}$, $a_{t|t}$, $\hat{y}_{t|t-1}$, and the sequences of conditional variance-covariance matrices, $\Sigma_{t|t-1}$, $\Sigma_{t|t}$, F_t. These quantities can be iteratively obtained by employing the Kalman filter, an algorithm whose equations are given as follows.[5]

Algorithm 5.1 (Kalman Filter)

- ***Step 1, Initialization***
  Set

$$a_{0|0} = \bar{a}_0, \quad \Sigma_{0|0} = \bar{P}_0,$$

  and set $t = 1$.
- ***Step 2, Prediction from*** $t-1$ ***to*** t
  $a_{t-1|t-1}$ *and* $\Sigma_{t-1|t-1}$ *are given, but* y_t *has not been observed yet. Compute*

$$a_{t|t-1} = T\, a_{t-1|t-1} + c \tag{5.20}$$
$$\Sigma_{t|t-1} = T\Sigma_{t-1|t-1}T' + Q \tag{5.21}$$
$$\hat{y}_{t|t-1} = M\, a_{t|t-1} + d \tag{5.22}$$
$$F_t = M\Sigma_{t|t-1}M' + H \tag{5.23}$$

- ***Step 3, Updating at*** t
  y_t *has been observed. Compute*

$$K_t = \Sigma_{t|t-1}M'F_t^{-1} \tag{5.24}$$
$$a_{t|t} = a_{t|t-1} + K_t(y_t - \hat{y}_{t|t-1}) \tag{5.25}$$
$$\Sigma_{t|t} = \Sigma_{t|t-1} - K_t M \Sigma_{t|t-1} \tag{5.26}$$

- ***Step 4***
  If $t < T$, *set* $t := t + 1$, *and go to Step 2; else, stop.*

The Kalman Filter delivers the sequence of means and variance-covariance matrices for the conditional distributions of interest. It is initialized by the mean $\bar{a}_0$ and variance $\bar{P}_0$ of the state distribution at time 0. In most cases, however, these will be unknown. We will now show how $\bar{a}_0$ and $\bar{P}_0$ can be suitably replaced.

For $t = 1$ we have

$$a_{1|0} = E(\alpha_1|\mathcal{Y}_0) = E(T\alpha_0 + c + \eta_1|\mathcal{Y}_0) = TE(\alpha_0|\mathcal{Y}_0) + c.$$

Since $\mathcal{Y}_0$ is a vector of constants, the expectation involved can be interpreted as being conditional on the trivial σ-algebra. If the parameters of the marginal distribution of α_0 are known, this expectation is just equal to the mean of that normal distribution, i.e. $a_{0|0} = E(\alpha_0|\mathcal{Y}_0) = \bar{a}_0$. If it is not known, it can be replaced by the unconditional mean of the transition equation (which is a

[5] The Kalman filter is due to [68], [69]. See [76] for a proof.

VAR(1) process) provided the latter is stationary. A similar argument can be given for the initial variance-covariance matrix. Thus, using the results in section B in the appendix, $a_{0|0}$ in algorithm 5.1 will be chosen as

$$a_{0|0} = (I - T)^{-1} c. \tag{5.27}$$

The elements of $\Sigma_{0|0}$, written in a column vector, are given by

$$vec(\Sigma_{0|0}) = (I - T \otimes T)^{-1} vec(Q). \tag{5.28}$$

If the transition equation is nonstationary, an initialization as described is not possible. In this case one may adopt the concept of a diffuse prior. The nonstationary case, however, is not relevant for our econometric analysis of term structure models, so it will not be discussed here.[6]

From the Kalman filter algorithm above, it is evident that the sequences $\{\Sigma_{t|t-1}\}$, $\{\Sigma_{t|t}\}$, and $\{F_t\}$ of variance-covariance matrices depend on system matrices and initial conditions but not on the observations $\mathcal{Y}_T$. It turns out that under suitable conditions, these matrices converge to some steady state values. If the eigenvalues of the matrix T are inside the unit circle, if Q and H are positive semidefinite, and if at least one of them is strictly positive definite, then the sequence $\{\Sigma_{t|t-1}, t = 1, \ldots, T\}$ will converge to a unique steady state matrix $\bar{\Sigma}_{-1}$ as T goes to infinity.[7] Moreover, $\bar{\Sigma}_{-1}$ is positive semidefinite, and this limit matrix is the same for any positive semidefinite initial variance-covariance matrix $\Sigma_{0|0}$.

If the Kalman filter equations (5.24) and (5.26) are inserted into (5.21) (led by one period) one obtains the matrix difference equation

$$\Sigma_{t+1|t} = T \left[\Sigma_{t|t-1} - \Sigma_{t|t-1} M' \left(M \Sigma_{t|t-1} M' + H \right)^{-1} M \Sigma_{t|t-1} \right] T' + Q \tag{5.29}$$

that the sequence $\{\Sigma_{t|t-1}\}$ has to satisfy. The steady state matrix $\bar{\Sigma}_{-1}$ is the fixed point of that difference equation, that is, it satisfies

$$\bar{\Sigma}_{-1} = T \left[\bar{\Sigma}_{-1} - \bar{\Sigma}_{-1} M' \left(M \bar{\Sigma}_{-1} M' + H \right)^{-1} M \bar{\Sigma}_{-1} \right] T' + Q \tag{5.30}$$

which is known as the algebraic Riccati equation.[8] For a scalar state vector, (5.30) can be solved analytically for $\bar{\Sigma}_{-1}$, for a vector-valued state, one has to rely on numerical methods.[9]

If $\Sigma_{t|t-1}$ converges, then the sequences of the F_t, K_t and $\Sigma_{t|t}$ also converge to steady state values $\bar{F}$, $\bar{K}$ and $\bar{\Sigma}$ which can be obtained from the Kalman filter equations. We have

---

6 See [46] or [59] for a thorough discussion of the initialization problem. [46] also treat the situation where only some of the parameters of the initial distribution are unknown.

7 See [57] for this proposition.

8 The fixed point of this difference equation does not need to be unique. However, it is unique under the stated conditions.

9 See [7].

$$\bar{F} = M\bar{\Sigma}_{-1}M' + H, \tag{5.31}$$
$$\bar{K} = \bar{\Sigma}_{-1}M'\bar{F}^{-1}, \tag{5.32}$$
$$\bar{\Sigma} = \bar{\Sigma} - \bar{K}M\bar{\Sigma}_{-1}. \tag{5.33}$$

The convergence properties of these quantities can be exploited for saving up computational time when running the Kalman filter. After each step of the Kalman filter one may check if some norm of $\Sigma_{t+1|t} - \Sigma_{t|t-1}$ is smaller than some prespecified criterion. If this is the case, one will switch to the steady state values, and in the subsequent iterations of the Kalman filter, equations (5.21), (5.23), (5.24) and (5.26) can be ignored.

The linear Gaussian state space model can be considered as a benchmark case for the statistical framework considered in this book. We will now give a short comment on what happens when the Kalman Filter is applied to a model with linear measurement and transition equation for which the errors are not Gaussian.

As explained by [60] the Kalman Filter outputs $a_{t|t-1}$, $\hat{y}_{t|t-1}$, and $a_{t|t}$ preserve an optimality property: they are the linear projections of α_t and y_t on $\mathcal{Y}_{t-1}$ and $\mathcal{Y}_t$, respectively. Hence, they are estimators which have smallest mean square errors in the restricted class of all linear estimators. However, they are not conditional expectations any more, since in the non-Gaussian case, the conditional expectations function is generally nonlinear in the conditioning variables. Correspondingly, $\Sigma_{t|t-1}$, F_t, and $\Sigma_{t|t}$ are the MSE matrices of $a_{t|t-1}$, $\hat{y}_{t|t-1}$, and $a_{t|t}$ respectively, but they lose their interpretation as conditional variance-covariance matrices of the states.

The Kalman filter generates three sequences of estimates: $\{a_{t|t-1}\}$, $\{\hat{y}_{t|t-1}\}$, and $\{a_{t|t}\}$. Based on these results two other problems can be solved. The first is that of making predictions that are more than one step ahead. That is, for a given time s one wants to construct $a_{t|s}$ and $\hat{y}_{t|s}$ for $t = s+2, s+3, \ldots$. The second problem is that of smoothing[10], i.e. the computation of the sequence $\{a_{t|T},\, t = 1, \ldots, T\}$. The general purpose of smoothing is to obtain estimators that have smaller MSEs than those obtained by filtering.[11]

For the construction of multistep predictions, suppose that for some time s, $\mathcal{Y}_s$ is available and for some $l > 2$ an l-step ahead prediction, i.e. estimates $a_{s+l|s}$ and $\hat{y}_{s+l|s}$ as well as the corresponding MSE-matrices are to be found.[12] Starting with the state in period s and applying the transition equation l times, the state at time $s + l$ can be written as

---

[10] The literature distinguishes between three types of smoothers and corresponding algorithms, fixed point smoothing, fixed lag smoothing, and fixed interval smoothing. See [4]. For econometric applications, and in particular for the applications considered in this book only fixed interval smoothing is relevant.

[11] In a time invariant state space model the gain from smoothing as opposed to filtering will be larger the larger H is compared to Q, see [59].

[12] For the multistep prediction problem, see, e.g., [59] or [24].

$$\alpha_{s+l} = T^l \alpha_s + \left(\sum_{i=1}^{l} T^{i-1} \right) c + \sum_{i=1}^{l} T^{l-i} \eta_{s+i}.$$

In the Gaussian case, the minimum mean square estimate of that state is the conditional mean, with expectation taken at time s, thus

$$a_{s+l|s} := E(\alpha_{s+l}|\mathcal{Y}_s) = T^l a_{s|s} + \left(\sum_{i=1}^{l} T^{i-1} \right) c. \tag{5.34}$$

The MSE of that prediction, which is equal to the conditional variance-covariance matrix is then given as

$$\Sigma_{s+l|s} = T^l \Sigma_{s|s} T^{l\prime} + \sum_{i=1}^{l} T^{i-1} Q T^{i-1\prime}. \tag{5.35}$$

Note that $a_{s|s}$ and $\Sigma_{s|s}$ in these formulas are obtained from the Kalman filter.

With $a_{s+l|s}$ and $\Sigma_{s+l|s}$ at hand, the minimum mean square estimate of $\hat{y}_{s+l}$ and its MSE matrix can be computed as

$$\hat{y}_{s+l|s} = M a_{s+l|s} + d,$$

and

$$F^l_{s+l} = M \Sigma_{s+l|s} M' + H,$$

respectively. If the assumption of Gaussian disturbances is dropped, $a_{s+l|s}$ and $\hat{y}_{s+l|s}$ are the minimum mean square estimators from the class of all estimators that are linear in $\mathcal{Y}_s$.

For obtaining the sequence of smoothed estimates $\{a_{t|T}, t = 1, \ldots, T\}$ one first runs the Kalman filter and stores the resulting filtered estimates $\{a_{t|t}\}$ and MSE matrices $\{\Sigma_{t|t}\}$ as well as the one-step predictions $\{a_{t+1|t}\}$ and their MSE matrices $\{\Sigma_{t+1|t}\}$. Then the smoothed estimates are obtained working backwards: first $a_{T-1|T}$ and $\Sigma_{T-1|T}$ are computed, then $a_{T-2|T}$ and $\Sigma_{T-2|T}$, and so on. This is done according to the recursion:[13]

$$a_{t|T} = a_{t|t} + \Sigma^*_t (a_{t+1|T} - a_{t+1|t}) \tag{5.36}$$

$$\Sigma_{t|T} = \Sigma_{t|t} + \Sigma^*_t (\Sigma_{t+1|T} - \Sigma_{t+1|t}) \Sigma^{*\prime}_t \tag{5.37}$$

with

$$\Sigma^*_t = \Sigma_{t|t} T' \Sigma^{-1}_{t+1|t}.$$

Again, in a Gaussian model, $a_{t|T}$ and $\Sigma_{t|T}$ are the mean and variance of α_t conditional on $\mathcal{Y}_T$,

$$\alpha_t | \mathcal{Y}_T \sim N(a_{t|T}, \Sigma_{t|T}).$$

[13] A derivation of the smoothing recursion can be found in [56].

5.3.3 Maximum Likelihood Estimation

The latter described the filtering and prediction problem under the assumption of known system matrices. We will now assume that the system matrices contain unknown elements which are collected in the vector ψ. Maximum likelihood (ML) estimation of ψ is particularly simple in a linear Gaussian state space model, since the Kalman filter can be used to construct those quantities from which in turn the likelihood function is constructed.

Under our normality assumption, the distribution of y_t conditional on $\mathcal{Y}_{t-1}$ is the N-dimensional normal distribution with mean $\hat{y}_{t|t-1}$ and variance-covariance matrix F_t. Thus, the conditional density of y_t can be written as[14]

$$\begin{aligned} &p(y_t|\mathcal{Y}_{t-1};\psi) \\ &= \left[(2\pi)^{N/2}\sqrt{|F_t|}\right]^{-1} \cdot \exp[-1/2(y_t - \hat{y}_{t|t-1})' F_t^{-1}(y_t - \hat{y}_{t|t-1})]. \end{aligned}$$

Accordingly, the log-likelihood function becomes

$$\ln \mathcal{L}(\psi) = l(\psi) = -\frac{NT}{2}\log 2\pi - \frac{1}{2}\sum_{t=1}^{T}\log|F_t| - \frac{1}{2}\sum_{t=1}^{T} v_t' F_t^{-1} v_t. \tag{5.38}$$

This function can be maximized with respect to ψ using some numerical optimization procedure.

The function (5.38) only depends on the prediction errors $v_t = y_t - \hat{y}_{t|t-1}$ and their variance-covariance matrices F_t. Both in turn are outputs of the Kalman Filter. Based on these results, maximum likelihood estimation of ψ can be summarized in a stylized way as follows:

1. Choose a value for ψ say ψ_0.
2. Run the Kalman Filter and store the sequences $\{v_t(\psi_0)\}$ and $\{F_t(\psi_0)\}$.
3. Use $\{v_t(\psi_0)\}$ and $\{F_t(\psi_0)\}$ to compute the log-likelihood in (5.38).
4. Employ an optimization procedure that repeats steps 1. - 3. until a maximizer $\hat{\psi}$ of (5.38) has been found.

When using a numerical optimization procedure, the required gradients can be computed numerically by using finite differences. However, the optimization process may be considerably stabilized by using analytical gradients. The ith element of the score vector is given by:[15]

$$\frac{\partial l(\psi)}{\partial \psi_i} = -\frac{1}{2}\sum_{t=1}^{T}\left[tr\left(\left(F_t^{-1}\frac{\partial F_t}{\partial \psi_i}\right)(I - F_t^{-1}v_t v_t')\right) - \frac{\partial v_t'}{\partial \psi_i}F_t^{-1}v_t\right]. \tag{5.39}$$

The required sequences of derivatives, $\{\frac{\partial v_t'}{\partial \psi_i}\}$ and $\{\frac{\partial F_t}{\partial \psi_i}\}$ can be computed using an iterative algorithm that can be run alongside the Kalman filter. For

[14] See, e.g., [60].
[15] See [59].

practical applications one usually faces a trade off between the effort that has to be spent on computing analytical gradients and the improvement in numerical stability.

Under certain conditions the maximum likelihood estimator $\hat{\psi}$ is asymptotically normally distributed,

$$\sqrt{T}(\hat{\psi} - \psi^0) \xrightarrow{d} N(0, \mathcal{I}_{HT}^{-1}) \tag{5.40}$$

where ψ^0 denotes the true value of the parameter vector and $\mathcal{I}_{HT}$ is the information matrix,

$$\mathcal{I}_{HT} = -\frac{1}{T} E\left(\left. \frac{\partial^2 l(\psi)}{\partial \psi \partial \psi'} \right|_{\psi=\psi^0} \right). \tag{5.41}$$

The regularity conditions for this asymptotic distribution to be valid include the following: the true parameter vector ψ^0 lies in the interior of the parameter space, the transition equation is stationary, and the parameters are identifiable.[16]

The information matrix can be estimated consistently by

$$\hat{\mathcal{I}}_{HT} = -\frac{1}{T} \left. \frac{\partial^2 l(\psi)}{\partial \psi \partial \psi'} \right|_{\psi=\hat{\psi}} = -\frac{1}{T} \sum_{t=1}^{T} \left. \frac{\partial^2 \ln p(y_t | \mathcal{Y}_{t-1}; \psi)}{\partial \psi \partial \psi'} \right|_{\psi=\hat{\psi}}. \tag{5.42}$$

Accordingly, the estimated variance-covariance matrix of $\hat{\psi}$ reported in empirical studies is given by

$$\hat{Var}(\hat{\psi}) = \frac{1}{T} (\hat{\mathcal{I}}_{HT})^{-1}, \tag{5.43}$$

i.e. it is the negative of the inverse Hessian of the log-likelihood evaluated at its maximizer.

If in the linear model the state innovation and measurement error are not Gaussian, one can still obtain estimates of the model parameters by (falsely) assuming normality, computing the log-likelihood by means of the Kalman filter, and maximizing it with respect to ψ. This approach is known as quasi-maximum likelihood estimation. Under certain conditions it will still lead to consistent estimators which are asymptotically normally distributed.[17]

Instead of relying on asymptotic standard errors, the distribution of the parameter estimators and its moments may be obtained by using the bootstrap.[18] This approach may be preferable to employing asymptotic results, especially if the sample size is small or if the distribution of the innovation or the measurement error is not normal. The procedure is described in [100] and can be summarized as follows.

---

16 See [76] or [57].

17 See [57] and the references given therein.

18 For the principles underlying the bootstrap see [47].

The state space model is transformed to a representation which is called the 'innovations form'. This is given by

$$y_t = d + M a_{t|t-1} + v_t \tag{5.44}$$

$$a_{t+1|t} = T a_{t|t-1} + K_t v_t \tag{5.45}$$

The quantities $a_{t|t-1}$, K_t and v_t, are the one-step prediction of the state vector, the Kalman gain matrix, and the one-step prediction error of the measurement vector, respectively. Note that this representation only contains the innovations v_t whereas the original representation of the state space model contained two sources of randomness, η_t and ϵ_t. The system (5.44) - (5.45) is again rewritten in terms of standardized innovations $\tilde{v}_t = F_t^{-1/2} v_t$ as

$$y_t = d + M a_{t|t-1} + F_t^{-1/2} \tilde{v}_t \tag{5.46}$$

$$a_{t+1|t} = T a_{t|t-1} + K_t F_t^{-1/2} \tilde{v}_t \tag{5.47}$$

The estimate $\hat{\psi}$ of the vector of unknown parameters is obtained by (quasi) maximum likelihood. Based on $\hat{\psi}$ the Kalman filter is run and a sequence of residual vectors $\{v_1(\hat{\psi}), \ldots, v_T(\hat{\psi})\}$ is obtained. Using the sequence of residual variance-covariance matrices $\{F_1(\hat{\psi}), \ldots, F_T(\hat{\psi})\}$, the series of standardized residuals $\tilde{v} = \{\tilde{v}_1(\hat{\psi}), \ldots, \tilde{v}_T(\hat{\psi})\}$ is obtained.

For the ith run within the bootstrap, one draws T times from $\tilde{v}$ with replacement and obtains a so-called bootstrap sample $\tilde{v}^{*i} = \{\tilde{v}_1^{*i}(\hat{\psi}), \ldots, \tilde{v}_T^{*i}(\hat{\psi})\}$ of standardized residuals. These are used in (5.46) - (5.47) to construct a bootstrap sample of observations $\mathcal{Y}_T^{*i} = \{y_t^{*i}, \ldots, y_T^{*i}\}$. Based on $\mathcal{Y}_T^{*i}$ the likelihood $\mathcal{L}(\psi|\mathcal{Y}_T^{*i})$ is constructed and maximized. Its maximizer $\hat{\psi}^{*i}$ is stored.

After C runs, e.g. $C = 1000$, a set of C bootstrap estimates $\{\hat{\psi}^{*1}, \ldots, \hat{\psi}^{*C}\}$ is obtained. The empirical distribution of $\{\hat{\psi}^{*1} - \hat{\psi}, \ldots, \hat{\psi}^{*C} - \hat{\psi}\}$ then serves as an approximation of the distribution of $\hat{\psi} - \psi^0$ where ψ_0 is the true parameter vector.

The advantage of using standardized residuals lies in the fact that they have the same variance-covariance matrix. However, we have seen above that the matrix F_t of the Kalman filter converges under certain circumstances to some steady-state value. In practice this steady state is usually reached after a few (less than ten) observations. Whether convergence has been reached can easily be checked by plotting the sequence of elements of F_t against time. Thus, an alternative to using standardized residuals is resampling from the set of raw residuals after the first few of them have been deleted. Construction of bootstrap observations y_t^{*i} is then based on (5.44) - (5.45).

The bootstrap approach just described may be referred to as nonparametric, since it imposes no distributional assumptions for generating the bootstrap observations y_t^{*i}. If one is confident that the model's state innovation and measurement error are in fact normal (or of some other 'known' distribution) one may resample from the state space model in its original form.

Bootstrap sequences of η_t and ϵ_t are generated by drawing from $N(0, Q(\hat{\psi}))$ and $N(0, H(\hat{\psi}))$ respectively. This approach would then be referred to as a parametric bootstrap.

6

State Space Models with a Gaussian Mixture

We consider the linear state space model and introduce a modification concerning the distribution of the state innovation. Instead of assuming normality, the innovation distribution is now specified to be a mixture of B normal distributions. In chapter 9 this type of state space model will be used for estimating a term structure model from the AMGM class. However, as the literature shows,[1] state space models involving mixture distributions have a variety of fields of application, especially in the engineering and natural sciences. Similar to the preceding chapter we will give a presentation of the material which is uncoupled from particular applications.

6.1 The Model

Although only one assumption is changed in comparison to the linear Gaussian model we write the model equations again for convenience. The transition equation is given by

$$\alpha_t = T\alpha_{t-1} + c + \eta_t, \tag{6.1}$$

where now for the innovation vector η_t

$$\eta_t \sim i.i.d. \sum_{b=1}^{B} \omega_b N(\mu_b, Q_b), \quad \sum_{b=1}^{B} \omega_b = 1, \quad \sum_{b=1}^{B} \omega_b \mu_b = 0. \tag{6.2}$$

That is, the density of η_t is given by[2]

$$p(\eta_t) = \sum_{b=1}^{B} \omega_b \phi(\eta_t; \mu_b, Q_b). \tag{6.3}$$

[1] See the references given in section 6.4 below.

[2] Recall that $\phi(x; \mu, Q)$ denotes the density function of $N(\mu, Q)$ evaluated at x.

This distribution has been introduced in section 3.2.2 and we will sometimes refer to the results given in that section.

For the variance-covariance matrix of η_t we have

$$Var(\eta_t) = \sum_{b=1}^{B} \omega_b \, (Q_b + \mu_b \, \mu_b') =: Q. \tag{6.4}$$

The measurement equation is again

$$y_t = M\alpha_t + d + \epsilon_t, \tag{6.5}$$

and the measurement error is still normally distributed,

$$\epsilon_t \sim i.i.d. \, N(0, H). \tag{6.6}$$

The measurement error ϵ_t and the state innovation η_s are independent for all times s and t. The weights ω_b as well as the system matrices and vectors T, c, M, d, H, μ_b, and Q_b are all assumed to be time-invariant. The initial state is also assumed to be normally distributed,

$$\alpha_0 \sim N(\bar{a}_0, \bar{P}_0), \tag{6.7}$$

and both, η_t and ϵ_t are independent from the initial state for all t.

Obviously, the model reduces to the standard linear Gaussian state space model if $B = 1$, or if all pairs (μ_b, Q_b) are identical. Replacing the normal distribution in the state process by a Gaussian mixture can be motivated by the fact that a wide variety of density functions can be approximated by a mixture of normals. Unlike for the case of a single normal, the mixture parameters ω_b, μ_b and Q_b can be chosen to determine the higher moments of the distribution independently from each other. This enables, for instance, to generate distributions which are heavy tailed, multimodal or asymmetric. As such, the specification considered here introduces more flexibility compared to the standard Gaussian model.

Finally, we want to introduce one possible extension of the mixture state space model in which the evolution of the innovation η_t is governed by an underlying Markovian indicator process.

Let I_t denote an independently, identically distributed discrete random variable that takes on values in $\mathcal{B} = \{1, \ldots, B\}$ with respective probabilities $\omega_1, \ldots, \omega_B$. Let $\mathcal{I}_t = \{I_1, \ldots, I_t\}$ be a trajectory of realizations of the indicator variable. Thus, the possible realizations of $\mathcal{I}_t$ are given by the set $\mathcal{B}_t := \mathcal{B}^t$. Using I_t we can give an alternative representation of the evolution of η_t as follows:

$$\begin{aligned} &I_t \text{ is an i.i.d. discrete random variable} \\ &\quad \text{with } P(I_t = b) = \omega_b, \, b = 1, \ldots, B. \end{aligned} \tag{6.8}$$

$$\begin{aligned} &\eta_t \sim N(\mu_t, Q_t), \quad \mu_t = \textstyle\sum_{b=1}^{B} \delta_{bt} \mu_b, \; Q_t = \sum_{b=1}^{B} \delta_{bt} Q_b \\ &\quad \text{with } \delta_{bt} = 1 \text{ if } I_t = b, \text{ and } \delta_{bt} = 0 \text{ otherwise.} \end{aligned} \tag{6.9}$$

That is, for the description of the state space model, (6.2) is replaced by (6.8) - (6.9). Thus, *conditional* on a sequence $\mathcal{I}_T \in \mathcal{B}_T$, the model is a standard linear Gaussian state space model, but with a time dependent distribution of η_t.

The latter representation opens the way to a generalization in which I_t follows a Markov-switching process. That is, all components of the model remain the same with the exception of (6.8) which is replaced by the following assumption:

I_t is a Markov-switching variable with transition probabilities given by

$$\omega = \begin{pmatrix} \omega_{11} & \omega_{21} & \dots & \omega_{B1} \\ \omega_{12} & \omega_{22} & \dots & \omega_{B2} \\ \vdots & \vdots & \ddots & \vdots \\ \omega_{1B} & \omega_{2B} & \dots & \omega_{BB} \end{pmatrix}$$

$$\text{where} \quad \omega_{ij} = P(I_t = j | I_{t-1} = i), \quad \sum_{j=1}^{B} \omega_{ij} = 1, \quad \text{for all } i. \tag{6.10}$$

We may sometimes refer to this generalization of the mixture state space model. The subsequent analysis and the application of the framework for estimating term structure models, however, will be based on the specification (6.2) for η_t.

State space models involving Gaussian mixture distributions show up in different guises in the literature. One of the earliest examples is [102] who consider a model with scalar state and measurement equation, in which the state innovation, the measurement error, and the initial density of the state are allowed to be mixtures of normals. It is extended to the case with a nonlinear transition and measurement equation by [3]. In a Bayesian context, [58] analyze a state space model where both the variance-covariance matrix H of the measurement error and the variance-covariance matrix Q of the state innovation depend on an indicator process. The model is referred to as the multi-state dynamic linear model and nests the model presented here. Other research dealing with state space models with Gaussian mixture distributions includes [108], [90], [73], [75], [98], [32], and [50]. The mixture is used to model the distribution of the measurement error, the state innovation or both.

The extension of our state space model in which the indicator variable is allowed to follow a hidden Markov process is a special case of the specification introduced by [71].[3] It is called the *dynamic linear model with Markov-switching* and allows all system matrices to be dependent on a Markov indicator process. This model nests previous approaches employing Markov-switching in state space models, as for instance [1], [2], and [112].

[3] See also [72].

6.2 The Exact Filter

We first assume that the system matrices are known and present the exact solution to the filtering problem and the one-step-prediction problem. Let $a_{t|t-1}$, $\hat{y}_{t|t-1}$ and $a_{t|t}$ denote the conditional expectations corresponding to the conditional densities $p(\alpha_t|\mathcal{Y}_{t-1})$, $p(Y_t|\mathcal{Y}_{t-1})$ and $p(\alpha_t|\mathcal{Y}_t)$, and denote by $\Sigma_{t|t-1}$, F_t and $\Sigma_{t|t}$ the corresponding variance-covariance matrices. It turns out that for the mixture model, the filtering and prediction densities can be generated in an iterative fashion. They are all mixtures of normals, with the number of components increasing exponentially with time. The relationships between filtering and prediction densities are given by the following theorems.[4]

Theorem 6.1 (Prediction density for the mixture model). *Let the filtering density at time $t-1$, $t = 1, 2, \ldots, T$, be given by a Gaussian mixture with l_{t-1} components,*

$$p(\alpha_{t-1}|\mathcal{Y}_{t-1}) = \sum_{i=1}^{l_{t-1}} \omega_{i,\,t-1|t-1} \cdot \phi(\alpha_{t-1};\, a_{i,\,t-1|t-1}, \Sigma_{i,\,t-1|t-1}).$$

Then the one-step-prediction density for the state is

$$\begin{aligned} &p(\alpha_t|\mathcal{Y}_{t-1}) \\ &= \sum_{b=1}^{B} \sum_{i=1}^{l_{t-1}} \omega_{bi,\,t|t-1}\, \phi(\alpha_t;\, a_{bi,\,t|t-1}, \Sigma_{bi,\,t|t-1}) \end{aligned} \tag{6.11}$$

with

$$\omega_{bi,\,t|t-1} = \omega_b\, \omega_{i,\,t-1|t-1}, \tag{6.12}$$

$$a_{bi,\,t|t-1} = T a_{i,\,t-1|t-1} + c + \mu_b, \tag{6.13}$$

$$\Sigma_{bi,\,t|t-1} = T \Sigma_{i,\,t-1|t-1} T' + Q_b. \tag{6.14}$$

After reindexing and setting $l_t = B \cdot l_{t-1}$ the prediction density can be written as

$$p(\alpha_t|\mathcal{Y}_{t-1}) = \sum_{i=1}^{l_t} \omega_{i,\,t|t-1} \phi(\alpha_t;\, a_{i,\,t|t-1}, \Sigma_{i,\,t|t-1}). \tag{6.15}$$

The one-step-prediction density for the observation vector is

$$p(y_t|\mathcal{Y}_{t-1}) = \sum_{i=1}^{l_t} \omega_{i,\,t|t-1}\, \phi(y_t;\, \hat{y}_{i,\,t|t-1}, F_{i,\,t}) \tag{6.16}$$

with

$$\hat{y}_{i,\,t|t-1} = M a_{i,\,t|t-1} + d, \tag{6.17}$$

$$F_{i,\,t} = M \Sigma_{i,\,t|t-1} M' + H. \tag{6.18}$$

[4] The earliest derivation of these relations for the case of scalar measurement and transition equation may be attributed to [102].

Proof. Making use of the general result (5.6) above, we have

$$
\begin{aligned}
&p(\alpha_t|\mathcal{Y}_{t-1})\\
&= \int p(\alpha_t|\alpha_{t-1})p(\alpha_{t-1}|\mathcal{Y}_{t-1})d\alpha_{t-1}\\
&= \int \left[\sum_{b=1}^{B} \omega_b \phi(\alpha_t;\, T\alpha_{t-1} + c + \mu_b, Q_b)\right]\\
&\quad \cdot \left[\sum_{i=1}^{l_{t-1}} \omega_{i,\,t-1|t-1}\phi(\alpha_{t-1};\, a_{i,\,t-1|t-1}, \Sigma_{i,\,t-1|t-1})\right] d\alpha_{t-1}\\
&= \sum_{b=1}^{B}\sum_{i=1}^{l_{t-1}} \omega_b\omega_{i,\,t|t-1} \int \phi(\alpha_t;\, T\alpha_{t-1} + c + \mu_b, Q_b)\\
&\quad \cdot\phi(\alpha_{t-1};\, a_{i,\,t-1|t-1}, \Sigma_{i,\,t-1|t-1})d\alpha_{t-1}
\end{aligned}
$$

For computing the integral of the two normal densities, lemma A.2 from the appendix is employed which leads directly to (6.11).

For the prediction density of the observation vector we have

$$
\begin{aligned}
&p(y_t|\mathcal{Y}_{t-1})\\
&= \int p(y_t|\alpha_t)p(\alpha_t|\mathcal{Y}_{t-1})d\alpha_t\\
&= \int \phi(y_t; M\alpha_t + d, H)\cdot\left(\sum_{i=1}^{l_t} \omega_{i,\,t|t-1}\phi(\alpha_t;\, a_{i,\,t|t-1}, \Sigma_{i,\,t|t-1})\right) d\alpha_t\\
&= \sum_{i=1}^{l_t} \omega_{i,\,t|t-1} \int \phi(y_t; M\alpha_t + d, H)\cdot\phi(\alpha_t;\, a_{i,\,t|t-1}, \Sigma_{i,\,t|t-1})d\alpha_t.
\end{aligned}
$$

Again applying lemma A.2 from the appendix yields the proposed density. □

Theorem 6.2 (Filtering density for the mixture model). *Let the prediction densities $p(\alpha_t|\mathcal{Y}_{t-1})$ and $p(y_t|\mathcal{Y}_{t-1})$ at time t, $t = 1, 2, \ldots, T$, be given by the Gaussian mixtures (6.15) and (6.16). Then the filtering density is*

$$
p(\alpha_t|\mathcal{Y}_t) = \sum_{i=1}^{l_t} \omega_{i\,t|t}\, \phi(\alpha_t;\, a_{i,\,t|t}, \Sigma_{i,\,t|t}) \tag{6.19}
$$

with

$$
\begin{aligned}
a_{i,\,t|t} &= a_{i,\,t|t-1} + K_{i,\,t}(y_t - \hat{y}_{i,\,t|t-1}), &(6.20)\\
\Sigma_{i,\,t|t} &= \Sigma_{i,\,t|t-1} - K_{i,\,t}M\Sigma_{i,\,t|t-1}, &(6.21)\\
K_{i,\,t} &= \Sigma_{i,\,t|t-1}M'F_{i,\,t}^{-1}, &(6.22)\\
\omega_{i,\,t|t} &= \frac{\omega_{i,\,t|t-1}\, \phi(y_t;\, \hat{y}_{i,\,t|t-1}, F_{i,\,t})}{\sum_{i=1}^{l_t} \omega_{i,\,t|t-1}\, \phi(y_t;\, \hat{y}_{i,\,t|t-1}, F_{i,\,t})}. &(6.23)
\end{aligned}
$$

Proof. In order to prove the result, one has to show that the proposed density (6.19) satisfies

$$p(\alpha_t|\mathcal{Y}_t)\,p(y_t|\mathcal{Y}_{t-1}) = p(y_t|\alpha_t)\,p(\alpha_t|\mathcal{Y}_{t-1}),$$

which is just a rearrangement of the general result (5.8) above. Note that $p(y_t|\mathcal{Y}_{t-1})$ and $p(\alpha_t|\mathcal{Y}_{t-1})$ are given and $p(y_t|\alpha_t)$ is implied by the measurement equation. Written in full one has to show that

$$\left(\sum_{i=1}^{l_t} \omega_{i,\,t|t}\phi(\alpha_t;\, a_{i,\,t|t}, \Sigma_{i,\,t|t})\right)\left(\sum_{i=1}^{l_t} \omega_{i,\,t|t-1}\phi(y_t;\, \hat{y}_{i,\,t|t-1}, F_{i,\,t})\right)$$
$$= \phi(y_t;\, M\alpha_t + d, H)\left(\sum_{i=1}^{l_t} \omega_{i,\,t|t-1}\phi(\alpha_t;\, a_{i,\,t|t-1}, \Sigma_{i,\,t|t-1})\right)$$

Plugging in the proposed expression for $\omega_{i,\,t|t}$ leads to

$$\sum_{i=1}^{l_t} \omega_{i,\,t|t-1}\phi(y_t;\, \hat{y}_{i,\,t|t-1}, F_{i,\,t})\phi(\alpha_t;\, a_{i,\,t|t}, \Sigma_{i,\,t|t})$$
$$= \sum_{i=1}^{l_t} \omega_{i,\,t|t-1}\phi(y_t;\, M\alpha_t + d, H)\phi(\alpha_t;\, a_{i,\,t|t-1}, \Sigma_{i,\,t|t-1}) \qquad (6.24)$$

We next show that for each i the product of the two densities on the left hand side and the right hand side are equal. Applying lemma A.2 from the appendix, the density product on the right hand side can be written as one multivariate normal density:

$$\phi(y_t;\, M\alpha_t + d, H)\phi(\alpha_t;\, a_{i,\,t|t-1}, \Sigma_{i,\,t|t-1})$$
$$= \phi\left(\begin{pmatrix} y_t \\ \alpha_t \end{pmatrix};\, \begin{pmatrix} Ma_{i,\,t|t-1} + d \\ a_{i,\,t|t-1} \end{pmatrix}, \begin{pmatrix} M\Sigma_{i,\,t|t-1}M' + H & M\Sigma_{i,\,t|t-1} \\ \Sigma_{i,\,t|t-1}M' & \Sigma_{i,\,t|t-1} \end{pmatrix}\right)$$

This multivariate density can be written as a product of a conditional and a marginal density,

$$p\left(\begin{pmatrix} y_t \\ \alpha_t \end{pmatrix}\right) = p(\alpha_t|y_t)\cdot p(y_t).$$

Since the joint density is normal, the marginal and conditional density are also normal. The moments of the conditional density are given in lemma A.3 in the appendix. Noting that $Ma_{i,\,t|t-1} + d = \hat{y}_{i,\,t|t-1}$ and $M\Sigma_{i,\,t|t-1}M' + H = F_{i,\,t}$, we have

$$\phi(y_t;\, M\alpha_t + d, H)\phi(\alpha_t;\, a_{i,\,t|t-1}, \Sigma_{i,\,t|t-1})$$
$$= \phi(\alpha_t;\, a_{i,\,t|t-1} + \Sigma_{i,\,t|t-1}M'F_{i,\,t}^{-1}(y_t - \hat{y}_{i,\,t|t-1}),$$
$$\Sigma_{i,\,t|t-1} - \Sigma_{i,\,t|t-1}M'F_{i,\,t}^{-1}M\Sigma_{i,\,t|t-1})$$
$$\cdot\phi(y_t;\, \hat{y}_{i,\,t|t-1}, F_{i,\,t}), \qquad (6.25)$$

Observing that the moments of the conditional density correspond to the proposed expressions for $a_{i,\,t|t}$ and $\Sigma_{i,\,t|t}$ it turns out that (6.25) is just the product of the ith two densities of the left hand side of (6.24). □

A remark is in order that theorems 6.1 and 6.2 are in fact applicable to time $t = 1$. For the initial filtering density used in theorem 6.1 we have $p(\alpha_0|\mathcal{Y}_0) = p(\alpha_0|1_N) = p(\alpha_0)$. Thus, technically speaking, the filtering density is the density of the initial state, that has been specified in (6.7) as a simple normal. It can be written as a mixture with one component, $l_0 = 1$, thus

$$p(\alpha_0|\mathcal{Y}_0) = \sum_{i=1}^{l_0} 1\, \phi(\alpha_0;\, \bar{a}_0, \bar{P}_0).$$

Hence, theorem 6.1 can be applied to this density yielding $p(\alpha_1|\mathcal{Y}_0)$ and $p(y_1|\mathcal{Y}_0)$ as mixtures with B components. To these in turn, theorem 6.2 can be applied yielding $p(\alpha_1|\mathcal{Y}_1)$.

It follows from the theorems above that filtering and prediction densities are mixtures of normals for all t. The number of components of these mixtures, however, increases exponentially in time, yielding $l_t = B^t$ components at time t. Thus, the computational cost of computing the exact densities also increases exponentially which makes it difficult to apply the exact filter to practical problems. Approximating the exact filter will be the subject of section 6.3.

With the conditional densities at hand, point estimators can be readily computed as the corresponding conditional expectations,

$$E(\alpha_t|\mathcal{Y}_{t-1}) = \sum_{i=1}^{l_t} \omega_{i,\,t|t-1}\, a_{i,\,t|t-1} =: a_{t|t-1}, \tag{6.26}$$

$$E(y_t|\mathcal{Y}_{t-1}) = \sum_{i=1}^{l_t} \omega_{i,\,t|t-1}\, \hat{y}_{i,\,t|t-1} =: \hat{y}_{t|t-1}, \tag{6.27}$$

$$E(\alpha_t|\mathcal{Y}_t) = \sum_{i=1}^{l_t} \omega_{i,\,t|t}\, a_{i,\,t|t} =: a_{t|t}. \tag{6.28}$$

The corresponding conditional variance-covariance matrices are given by

$$Var(\alpha_t|\mathcal{Y}_{t-1}) = \sum_{i=1}^{l_t} \omega_{i,\,t|t-1} \left(\Sigma_{i,\,t|t-1} + (a_{i,\,t|t-1} - a_{t|t-1})(a_{i,\,t|t-1} - a_{t|t-1})'\right)$$
$$=: \Sigma_{t|t-1} \tag{6.29}$$

$$Var(y_t|\mathcal{Y}_{t-1}) = \sum_{i=1}^{l_t} \omega_{i,\,t|t-1} \left(F_{i,\,t} + (\hat{y}_{i,\,t|t-1} - \hat{y}_{t|t-1})(\hat{y}_{i,\,t|t-1} - \hat{y}_{t|t-1})'\right)$$
$$=: F_{t|t-1} \tag{6.30}$$

$$Var(\alpha_t|\mathcal{Y}_t) = \sum_{i=1}^{l_t} \omega_{i,\,t|t} \left(\Sigma_{i,\,t|t} + (a_{i,\,t|t} - a_{t|t})(a_{i,\,t|t} - a_{t|t})'\right)$$
$$=: \Sigma_{t|t}. \tag{6.31}$$

The latter results follow from the general properties of Gaussian mixtures presented in section 3.2.2 above. Note that the expectation is just the weighted average of the expectations of the normal densities that constitute the mixture, whereas the variance has an additional term taking the variation of the means into account.

It is instructive to compare the filtering and prediction densities for the mixture model with the Kalman filter. We first note that the Kalman filter yields best linear estimators for the mixture model under consideration. This simply follows from the fact that the model possesses a linear state and measurement equation. However, due to the non-Gaussian distribution of the error in the transition equation, the best linear estimator is not the best estimator overall. In fact, the optimal estimators, i.e. the conditional expectations $a_{t|t}$, $a_{t|t-1}$ and $\hat{y}_{t|t-1}$ computed by (6.26) - (6.28), are not linear in $\mathcal{Y}_t$. This is observed by noting that the weights, computed in (6.12) and (6.23), depend on $\mathcal{Y}_t$ in a nonlinear fashion. Moreover, unlike in case of a linear Gaussian model, the conditional variances $\Sigma_{t|t}$, $\Sigma_{t|t-1}$ and F_t do depend on the observations $\mathcal{Y}_t$.

If we set $B = 1$, the model reduces to the standard linear Gaussian state space model as remarked in the previous section. For the case $B = 1$, it is easily observed that the equations for computing the filtering and prediction densities in theorems 6.1 and 6.2 coincide with those of the Kalman filter. For $B \geq 2$, however, the operations that compute the components of the respective mixture distributions can be interpreted as a bunch of Kalman filters working parallel. For fixed i and b, for example, the prediction equations have the form of the usual Kalman filter equation for constructing the state predictor and its variance-covariance matrix. However, since the operation has to be conducted for all b and i, the Kalman filter prediction step has to be made $l_t = B^t$ times. A similar comment applies to the updating equations.

If $B > 1$ but $Q_b = Q$ and $\mu_b = 0$ for all b, the density of the state innovation η_t is a simple normal. It may be said to have a 'blown up' representation, since one mixes identical component densities. Accordingly, the exact filtering and prediction densities given in theorems 6.1 and 6.2 will be Gaussian. This is intuitive and can be seen as follows. If $Q_b = Q$ and $\mu_b = 0$, then at each time t, all component means $a_{i,\,t|t-1}$ are the same. The same is true for $\hat{y}_{i,\,t|t-1}$ and $a_{i,\,t|t}$, and the component variance-covariance matrices $\Sigma_{i,\,t|t-1}$, $F_{i,\,t}$ and $\Sigma_{i,\,t|t}$. This can be easily checked by going through the equations in theorems 6.1 and 6.2. The weights $\omega_{i,\,t|t}$ and $\omega_{i\,t|t-1}$ do not depend on the observations in this special case, since the densities involved in (6.23) cancel. Thus, if one were to compute the filtering and prediction densities according to theorems 6.1 and 6.2, one would obtain 'degenerated' mixtures of normals: for each mixture, the component densities are all the same, thus the mixture density function is identical to a simple normal density function. This implies that for the case $Q_b = Q$ and $\mu_b = 0$, the outputs of the exact filter coincide with those of the Kalman filter. That is, both filters generate the same sequences of means $\{a_{t|t-1}\}$, $\{\hat{y}_{t|t-1}\}$, $\{a_{t|t}\}$, and variance-covariance matrices, $\{\Sigma_{t|t-1}\}$, $\{F_t\}$, $\{\Sigma_{t|t}\}$.

We will now summarize the steps of the exact filter for the mixture model. Given observations $\{y_1, \ldots, y_T\}$, and an initial density $\alpha_0 \sim N(\bar{a}_0, \bar{P}_0)$, the algorithm computes

- the sequences of conditional densities,

$$
\begin{aligned}
&p(\alpha_t|\mathcal{Y}_{t-1}), \quad t = 1, \ldots, T,\\
&p(y|\mathcal{Y}_{t-1}), \quad t = 1, \ldots, T,\\
&p(\alpha_t|\mathcal{Y}_t), \quad t = 1, \ldots, T,
\end{aligned}
$$

  each characterized by the corresponding components (weights, means, variances),

$$
\begin{aligned}
&\omega_{i,\,t|t-1}, a_{i,\,t|t-1}, \Sigma_{i,\,t|t-1}, \quad i = 1, \ldots, l_t \quad t = 1, \ldots, T,\\
&\omega_{i,\,t|t-1}, \hat{y}_{i,\,t|t-1}, F_{i,\,t}, \quad i = 1, \ldots, l_t \quad t = 1, \ldots, T,\\
&\omega_{i,\,t|t}, a_{i,\,t|t}, \Sigma_{i,\,t|t}, \quad i = 1, \ldots, l_t \quad t = 1, \ldots, T,
\end{aligned}
$$

- and the sequences of point estimates (conditional means) and corresponding variance-covariance matrices

$$
\begin{aligned}
&a_{t|t-1}, \Sigma_{t|t-1}, \quad t = 1, \ldots, T,\\
&\hat{y}_{t|t-1}, F_t, \quad t = 1, \ldots, T,\\
&a_{t|t}, \Sigma_{t|t}, \quad t - 1, \ldots, T.
\end{aligned}
$$

  These are computed according to the following scheme:

Algorithm 6.1 (The exact filter)

- ***Step 1, Initialization***
  Set

$$
a_{1,\,0|0} = \bar{a}_0, \quad \Sigma_{1,\,0|0} = \bar{P}_0, \quad \omega_{1,\,0|0} = 1, \quad l_0 = 1.
$$

  Set $t = 1$.

- ***Step 2, Prediction step from*** $t-1$ ***to*** t
  Set $l_t = B^t$.
  Compute $\omega_{i,\,t|t-1}$, $a_{i,\,t|t-1}$, $\Sigma_{i,\,t|t-1}$, $\hat{y}_{i,\,t|t-1}$, *and* $F_{i,\,t}$ *for* $i = 1, \ldots, l_t$, *according to theorem 6.1.*
  Use these quantities to compute $a_{t|t-1}$, $\hat{y}_{t|t-1}$, $\Sigma_{t|t-1}$, *, and* F_t *according to (6.26), (6.27), (6.29) and (6.30) respectively.*
- ***Step 3, Updating step at*** t
  Compute $\omega_{i,\,t|t}$, $a_{i,\,t|t}$, *and* $\Sigma_{i,\,t|t}$, *for* $i = 1, \ldots, l_t$, *according to theorem 6.2.*
  Use theses quantities to compute $a_{t|t}$ *and* $\Sigma_{t|t}$, *according to (6.28) and (6.31), respectively.*
- ***Step 4***
  If $t < T$, *set* $t := t + 1$, *and go to Step 2;*
  else, STOP.

If the moments of the initial conditions are not known, one can proceed as in the case of a simple normal. If the state process is stationary, the filter can be initialized using (5.27) and (5.28). The condition for stationarity of the state process is the same as in the Gaussian case: all eigenvalues of the transition matrix T have to have modulus less than one.

As mentioned above, in economic applications of state space models system matrices generally depend on unknown parameters. Let them be collected in a vector ψ. We now turn to the problem of estimating ψ by maximum likelihood.

The likelihood of the data as a function of ψ is given by

$$\mathcal{L}(\psi) = p(y_1, \ldots, y_T; \psi) = \prod_{t=1}^{T} p(y_t|\mathcal{Y}_{t-1}; \psi). \tag{6.32}$$

The ML-estimator of ψ is

$$\hat{\psi}^{ML} = \arg\max_{\psi} \ln \mathcal{L}(\psi).$$

The likelihood $\mathcal{L}(\psi)$ is a product of the one-step-prediction densities for the observation vector. For a linear Gaussian state space model these are normal densities. For the mixture model, however, theorem 6.1 implies that $p(y_t|\mathcal{Y}_{t-1})$ is a mixture of $l_t = B^t$ Gaussian components. Hence, for this case the log-likelihood an be written as:

$$\begin{aligned} l(\psi) &= \ln \mathcal{L}(\psi) \\ &= \sum_{t=1}^{T} \ln p(y_t|\mathcal{Y}_{t-1}; \psi) \\ &= \sum_{t=1}^{T} \ln \left(\sum_{i=1}^{l_t} \omega_{i,\,t|t-1}\, \phi(y_t;\, \hat{y}_{i,\,t|t-1}, F_{i,\,t}) \right) \\ &= \sum_{t=1}^{T} \ln \left(\sum_{i=1}^{l_t} \frac{\omega_{i,\,t|t-1}}{(2\pi)^{N/2}\sqrt{|F_{i,\,t}|}}\, e^{-\frac{1}{2}(y_t - \hat{y}_{i,\,t|t-1})' F_{i,\,t}^{-1} (y_t - \hat{y}_{i,\,t|t-1})} \right) \end{aligned} \tag{6.33}$$

In the case of a linear Gaussian state space model, we have $l_t = 1$ for all t and the last expression simplifies. For the mixture model, the logarithm operates on a sum, however, preventing simplification.

For constructing the likelihood for a given choice of ψ, say ψ_0, one has to run algorithm 6.1 for computing the sequence $\{p(y_t|\mathcal{Y}_{t-1};\psi_0)\}$ of conditional densities. That implies that when searching the maximizer of (6.33) numerically, one has to run algorithm 6.1 for each required evaluation of the likelihood, i.e. for each trial value of ψ.

6.3 The Approximate Filter AMF(k)

In the previous subsection we have seen that the exact filtering and prediction densities are mixtures of normals, with the number of components growing exponentially. This implies that the exact filter cannot be applied for time series that contain more than a few observations. With $B = 2$, for instance, the exact filtering density at $t = 10$ already contains $2^{10} = 1024$ components. We propose an approximation scheme for which the maximum number of components appearing in the employed mixture distributions is governed by a parameter $k < T$. After an initial phase, the exact filtering and prediction densities, mixtures with B^t components, are approximated by mixtures with B^k components only. This approximating density results from applying the exact filter to the most recent k observations only. A suitable initialization of the filter takes the first $t-k$ observations into account in a condensed form. Next, we describe in detail how the approximation works.[5]

First, the exact filter is run up to time $t = k$ yielding the exact filtering densities $p(\alpha_t|\mathcal{Y}_t)$ for $t = 1, \ldots, k$ as described in the previous section. The last of these densities, $p(\alpha_k|\mathcal{Y}_k)$, is a mixture of B^k normals.

Continuing with the exact filter would deliver the exact density for time $t = k+1$ as a mixture with B^{k+1} components. However, we want to constrain the number of components to B^k. The idea is now to apply the exact filter algorithm, but only to the last k observations of $\mathcal{Y}_{k+1}$, i.e. to the subsequence $\{y_2, \ldots, y_{k+1}\}$. The filter is initialized by the univariate normal with mean $a_{1|1}$ and variance $\Sigma_{1|1}$, the latter being the mean and the variance of the B-component mixture $p(\alpha_1|\mathcal{Y}_1)$. Thus, the initial condition contains information about y_1 in a condensed form, the exact density $p(\alpha_1|\mathcal{Y}_1)$ is replaced by a simple normal. Applying the exact filter in this fashion to the most recent k observations yields a mixture with B^k components, denoted by $\tilde{p}(\alpha_{k+1}|\mathcal{Y}_{k+1})$, that approximates the exact filtering density at time $k+1$.

A similar procedure is applied for approximating each of the filtering densities from $t = k+1$ to $t = 2k$. For obtaining an approximation of the density

[5] We will refer to the filtering densities only. The idea is the same for the prediction densities. In the summary of the approximation algorithm below, it will be documented how they are computed.

$p(\alpha_t|\mathcal{Y}_t)$, the exact filter is applied to the k most recent observations only. The first $t-k$ observations $\{y_1,\ldots,y_{t-k}\}$, however, are not ignored. They enter the estimation process through the initial condition. The exact filter is initialized by a simple normal, and the mean of that normal is $a_{t-k|t-k}$, the optimal estimate of the state at $t-k$, given the observations from 1 to $t-k$. Since the algorithm is iteratively applied, the estimate $a_{t-k|t-k}$ and its variance-covariance matrix $\Sigma_{t-k|t-k}$ are already available.

In this fashion approximate densities $\tilde{p}(\alpha_t|\mathcal{Y}_t)$ for $t=k+1,\ldots,2k$ are obtained. Each of them is a mixture of B^k components.

Analogous operations can be conducted for approximating the filtering densities for $t=2k+1,\ldots,T$. At time $t\geq 2k+1$ the approximate density is generated by an application of the exact filter to $\{y_{t-k+1},\ldots,y_t\}$. For computing the initial condition at time $t-k$, one would again collapse the mixture density $p(\alpha_{t-k}|\mathcal{Y}_{t-k})$ to a simple normal. However, since we are beyond $t=2k$, we do not have the exact filtering density $p(\alpha_{t-k}|\mathcal{Y}_{t-k})$ for time $t-k$ available. We only have $\tilde{p}(\alpha_{t-k}|\mathcal{Y}_{t-k})$ available, a mixture of B^k components that approximates $p(\alpha_{t-k}|\mathcal{Y}_{t-k})$. Nevertheless, we can proceed as usual and collapse this density into a simple normal.

The approximation scheme will be summarized in algorithm 6.2. The filtering densities that are generated by the approximate filter will be denoted by $\tilde{p}(\alpha_t|\mathcal{Y}_t)$, $t=1,\ldots,T$.[6] We write $\tilde{a}_{t|t}$ and $\tilde{\Sigma}_{t|t}$ for the mean and variance implied by $\tilde{p}(\alpha_t|\mathcal{Y}_t)$. Alongside the approximate filtering densities, approximate prediction densities $\tilde{p}(\alpha_t|\mathcal{Y}_{t-1})$ and $\tilde{p}(y_t|\mathcal{Y}_{t-1})$ are also constructed. The corresponding means and variances are denoted by $\tilde{a}_{t|t-1}$, $\tilde{\hat{y}}_{t|t-1}$, $\tilde{\Sigma}_{t|t-1}$, and $\tilde{F}_t$ respectively. The components of the approximating mixture distributions are also denoted by a tilde. We thus write, e.g.,

$$\tilde{p}(\alpha_t|\mathcal{Y}_t)=\sum_{i=1}^{k}\tilde{\omega}_{i,\,t|t}\,\phi(\alpha_t;\,\tilde{a}_{i,\,t|t},\tilde{\Sigma}_{i,\,t|t}).$$

and similarly for $\tilde{p}(\alpha_t|\mathcal{Y}_{t-1})$ and $\tilde{p}(y_t|\mathcal{Y}_{t-1})$. We abbreviate the approximation scheme as AMF(k), standing for 'approximate mixture filter of degree k'.

Algorithm 6.2 (The approximate filter AMF(k))

- ***Step 1***
  Apply the exact filter to the sequence $\{y_1,\ldots,y_k\}$ with initial condition $\alpha_0\sim N(\bar{a}_0,\bar{P}_0)$.
  Obtain the exact filtering densities

$$p(\alpha_t|\mathcal{Y}_t),\quad p(\alpha_t|\mathcal{Y}_{t-1}),\quad p(y_t|\mathcal{Y}_{t-1}),\quad t=1,\ldots,k,$$

  with corresponding moments

$$a_{t|t},\ \Sigma_{t|t},\quad a_{t|t-1},\ \Sigma_{t|t-1},\quad \hat{y}_{t|t-1},\ F_t.$$

[6] Note, however, that for $t=1,\ldots,k$ these are exact as the approximate filter coincides with the exact filter for computing the first k conditional densities.

- ***Step 2***
  For $t = 1, \ldots, k$ set:

$$\begin{aligned} \tilde{p}(\alpha_t|\mathcal{Y}_t) &= p(\alpha_t|\mathcal{Y}_t), \quad \tilde{a}_{t|t} = a_{t|t}, \quad \tilde{\Sigma}_{t|t} = \Sigma_{t|t} \\ \tilde{p}(\alpha_t|\mathcal{Y}_{t-1}) &= p(\alpha_t|\mathcal{Y}_{t-1}), \quad \tilde{a}_{t|t-1} = a_{t|t-1}, \quad \tilde{\Sigma}_{t|t-1} = \Sigma_{t|t-1} \\ \tilde{p}(y_t|\mathcal{Y}_{t-1}) &= p(y_t|\mathcal{Y}_{t-1}), \quad \tilde{\hat{y}}_{t|t-1} = \hat{y}_{t|t-1}, \quad \tilde{F}_t = F_t. \end{aligned}$$

  Set $t = k + 1$.
- ***Step 3***
  Apply the exact filter to the sequence $\{y_{t-k+1}, \ldots, y_t\}$ with initial condition $\alpha_{t-k} \sim N(\tilde{a}_{t-k|t-k}, \tilde{\Sigma}_{t-k|t-k})$.
  Store the final filtering and prediction densities as $\tilde{p}(\alpha_t|\mathcal{Y}_t)$, $\tilde{p}(\alpha_t|\mathcal{Y}_{t-1})$, and $\tilde{p}(y_t|\mathcal{Y}_{t-1})$. That is, store the corresponding components $\tilde{\omega}_{i,\,t|t}$, $\tilde{a}_{i,\,t|t}$, $\tilde{\Sigma}_{i,\,t|t}$, $\tilde{\omega}_{i,\,t|t-1}$, $\tilde{a}_{i,\,t|t-1}$, $\tilde{\Sigma}_{i,\,t|t-1}$, $\tilde{\hat{y}}_{i,\,t|t-1}$, $\tilde{F}_{i,\,t}$, $\quad i = 1, \ldots, B^k$.
  Compute the corresponding means and variances $\tilde{a}_{t|t}$, $\tilde{\Sigma}_{t|t}$, $\tilde{a}_{t|t-1}$, $\tilde{\Sigma}_{t|t-1}$, $\tilde{\hat{y}}_{t|t-1}$, and $\tilde{F}_t$.
- ***Step 4***
  If $t < T$, set $t := t + 1$, and go to Step 3;
  else, STOP.

The AMF(k) coincides with the Kalman filter if $Q_b = Q$ and $\mu_b = 0$ for all b, a property which is inherited from the exact filter. Again, this is intuitive and can be explained as follows. For $t = 1, \ldots, k$ the outputs $\{\tilde{a}_{t|t}\}$ and $\{\tilde{\Sigma}_{t|t}\}$ of the AMF(k) are identical to those of the exact filter. As argued above, these in turn coincide with the outputs of the Kalman filter if $Q_b = Q$ and $\mu_b = 0$. For $t = k+1, \ldots, 2k$, the AMF(k) outputs result from applying the exact filter to $\{y_{t-k+1}, \ldots, y_t\}$ with initial condition based on $\tilde{a}_{t-k|t-k}$ and $\tilde{\Sigma}_{t-k|t-k}$. Now, again, the filter outputs $\tilde{a}_{t|t}$ and $\tilde{\Sigma}_{t|t}$ are identical to applying the Kalman filter to $\{y_{t-k+1}, \ldots, y_t\}$ with initial condition $\tilde{a}_{t-k|t-k}$ and $\tilde{\Sigma}_{t-k|t-k}$. Moreover, the initializing mean $\tilde{a}_{t-k|t-k}$ and variance-covariance matrix $\tilde{\Sigma}_{t-k|t-k}$ are identical to the outputs of the Kalman filter based on $\{y_1, \ldots, y_{t-k}\}$. Hence, $\tilde{a}_{t|t}$ and $\tilde{\Sigma}_{t|t}$ are identical to $a_{t|t}^{KF}$ and $\Sigma_{t|t}^{KF}$ that would be obtained by applying the Kalman filter to $\{y_1, \ldots, y_t\}$. The same argument goes through for $t > 2k$.

Of course, if one is faced with a model with $Q_b = Q$ and $\mu_b = 0$ for all b, then this is in fact a Gaussian model and the Kalman filter can be applied. Stating the latter result only served to show that the AMF(k) satisfies the consistency property of coinciding with the Kalman filter in that special case.

When collapsing the mixture to a single normal within our approximating algorithm, we choose the mean and the variance of the mixture as mean and variance for the approximating simple normal. This makes sense intuitively. In fact, the following theorem states that among all simple normals, the one with that particular mean and variance pair minimizes the Kullback-Leibler distance to the mixture.[7]

Theorem 6.3 (Optimal collapsing normal). *Let $p_1(x)$ be a mixture of l k-variate normal densities, $p_1(x) = \sum_{i=1}^{l} \omega_i \cdot \phi(x;\, \mu_i, V_i)$. Denote by $p_c(x) = \phi(x;\, \mu_c, V_c)$ that simple normal density which is closest in Kullback-Leibler distance to $p_1(x)$. Then μ_c and V_c satisfy*

$$\mu_c = \sum_{i=1}^{l} \omega_i \mu_i, \quad V_c = \sum_{i=1}^{l} \omega_i \left[V_i + (\mu_i - \mu)(\mu_i - \mu)' \right]. \tag{6.34}$$

Proof. See section F in the appendix. □

In order to make parameter estimation for the mixture model feasible, we propose the construction of an approximation to the exact likelihood in (6.33). This is obtained by replacing the conditional densities $p(y_t|\mathcal{Y}_{t-1})$ by their approximate counterparts $\tilde{p}(y_t|\mathcal{Y}_{t-1})$ from the AMF(k) . We write

$$\tilde{\mathcal{L}}(\psi) = \prod_{t=1}^{T} \tilde{p}(y_t|\mathcal{Y}_{t-1}) \tag{6.35}$$

for the approximate likelihood and

$$\begin{aligned} \tilde{l}(\psi) &= \sum_{t=1}^{T} \ln \tilde{p}(y_t|\mathcal{Y}_{t-1}) \\ &= \sum_{t=1}^{T} \ln \sum_{i=1}^{k} \tilde{\omega}_{i,\,t|t-1} \phi(y_t;\, \tilde{\bar{y}}_{i,\,t|t-1}, \tilde{F}_{i,\,t}) \end{aligned} \tag{6.36}$$

for the approximate log-likelihood. Approximate maximum likelihood estimates of the parameters are then given by

$$\hat{\psi} = \arg\max_{\psi} \ln \tilde{l}(\psi). \tag{6.37}$$

If parameter estimation is based on the exact likelihood, an estimate of the variance-covariance matrix of the parameter estimates could be based on the inverse Hessian of the log-likelihood at its maximum. That is, one could use formula (5.43) where l in (5.42) is now the exact likelihood of the mixture model. For estimates based on the approximate likelihood, we suggest using

[7] The case of a multivariate mixture with two components is proved in [90]. The result is generalized by Bolstad [21] who considers optimal collapsing of mixtures of two-dimensional exponential family members to a single member. His result is proved for an arbitrary number of components in the mixture but for univariate densities only. The proof given here (in the appendix) follows the lines of the proof in [90], extending it to an arbitrary number of components.

the same formulas, where the Hessian of the exact likelihood is replaced by the Hessian of the approximate likelihood. We leave the assessment of the quality of this approximation as a topic for future research.

Alternatively, the bootstrap can be employed to approximate the distribution of $\hat{\psi}$. The parametric version of the bootstrap would work analogously to the Gaussian case. Based on the estimated system parameters, one generates bootstrap observations $\mathcal{Y}_T^*$, from which bootstrap estimates of ψ are obtained. For simulating data from the mixture state space model, one has to generate innovations of the state process, which in turn requires generating pseudo random variates from the Gaussian mixture distribution.

Optimal multistep predictions in the mixture state space model can be constructed using the same approach as for the Gaussian model. Consider again the task of making a prediction of the state vector at time $t+s$ given observations $\mathcal{Y}_t$ up to time t. As in the Gaussian case, the future state vector can be written in terms of the current state and future innovations:

$$\alpha_{t+s} = T^s \alpha_s + \left(\sum_{i=1}^{s} T^{i-1} \right) c + \sum_{i=1}^{s} T^{s-i} \eta_{s+i}.$$

Hence, also for the mixture model, the MSE-optimal forecast is

$$E(\alpha_{t+s}|\mathcal{Y}_t) = T^s E(\alpha_t|\mathcal{Y}_t) + \left(\sum_{i=1}^{s} T^{i-1} \right) c.$$

Thus, based on the results of the exact filter, $a_{t+s|t} := E(\alpha_{t+s}|\mathcal{Y}_t)$ is given by

$$a_{t+s|t} = T^s a_{t|t} + \left(\sum_{i=1}^{s} T^{i-1} \right) c. \tag{6.38}$$

However, as explained above, computation of $a_{t|t}$ is computationally infeasible for larger t, so we replace it by its counterpart $\tilde{a}_{t|t}$ obtained from the AMF(k). This yields the formula for a multistep prediction based on the AMF(k),

$$\tilde{a}_{t+s|t} = T^s \tilde{a}_{t|t} + \left(\sum_{i=1}^{s} T^{i-1} \right) c. \tag{6.39}$$

By similar arguments as given for multistep predictions for the Gaussian state space model, the approximate MSE of that prediction is given by

$$\tilde{\Sigma}_{t+s|t} = T^s \tilde{\Sigma}_{t|t} T^{s\prime} + \sum_{i=1}^{s} T^{i-1} Q T^{i-1\prime}.$$

6.4 Related Literature

The literature that deals with state space models involving Gaussian mixtures or Markov-switching environments offers a wide variety of approaches for estimation. In most cases, only the problems of filtering, prediction and smoothing are of interest. Estimation of unknown parameters is treated seldom.[8]

The key problem to be tackled is the exponentially growing number of components in the prediction and filtering densities.

On a conceptual level this problem can be approached as follows. The true density is a mixture p_1 of L_1 normals, and it should be approximated by a mixture p_2 with $L_2 < L_1$ components. As a measure of distance one could use the Kullback-Leibler measure. Then one would have to solve the following optimization problem: given the mixture p_1 with L_1 components, find the values of the components (weights, means, variance-covariance matrices) of p_2, that minimize the distance from p_2 to p_1. Except for the trivial case with $L_2 = 1$ which is the subject of theorem 6.3 above, the solution to that problem requires some numerical effort that increases with L_2.

Another approach to reducing the number of components is to drop those components that have small corresponding weights in the mixture. Versions of this approach are employed by [102], [112] ("detection-estimation algorithm"), and [51]. [102] also suggest merging those components of which the means and variances are very similar to each other. The latter idea is also worked out by [73], [32], and [51] who pool two component densities of a mixture if the distance to each other is sufficiently small. The distance measures employed in these three articles, however, are different from each other.

Other articles use collapsing approaches similar to the AMF(k) . Within these approximate filters there are always steps in which mixtures are collapsed into a simple normal, see, e.g. [1], [75], and [50]. [58] suggest a collapsing scheme in which each time when the filter generates a mixture with B^2 components, this is reduced to B components. However, to our knowledge, the AMF(k) proposed here is unique with respect to the way in which the 'degree of approximation' can be flexibly controlled by the parameter k.

An alternative strategy is to use Monte Carlo methods for filtering, prediction and smoothing. To get the idea of the approach, consider again representation (6.8) - (6.9) of the evolution of the state innovation η_t. *Conditional* on a path of the indicator variable I_t, the model is a standard linear Gaussian state space model with time-dependent variance-covariance matrix Q_t. That is, given a particular trajectory of the indicator variable I_t, the filtering and prediction problem can be solved using the Kalman filter. According to the ω parameters, the Monte Carlo approach generates sequences $\{I_t\}$ at random. Then the Kalman filter results corresponding to each sequence are combined using a suitably weighted average. With N being the number of generated sequences of I_t, Akashi and Kumamoto [2, p. 433] interpret this approach as "learning with N probabilistic teachers". Monte Carlo methods are also employed by [112], [31], see also the exposition in [46].

---

[8] An important exception is Kim and Nelson [72] who show how approximate likelihood estimation can be conducted for their dynamic linear model with Markov-switching. Another example is [32].

For an intermediate conclusion concerning the AMF(k) proposed in this book, we can refer to three criteria proposed by [80]. They state that a filter for non-Gaussian situations has to be computationally attractive, easy to understand, and easy to implement. We think that the AMF(k) satisfies all three of these criteria. In practice one would have to decide over the parameter k. We propose that one starts with $k = 1$ and then moves on gradually to $k = 2$, $k = 3$ and so forth. If by stepping from k to $k + 1$ the results do not change much, one should stick with k. The simulation results of the next chapter[9] suggest that often the choice of $k = 1$ can hardly be improved upon. However, we also present circumstances under which it is worthwhile to use the AMF(k) with $k = 3$.

[9] This is also true for the empirical study in chapter 9.

7

Simulation Results for the Mixture Model

After discussing mixture state space models in the previous section, we now present a small simulation study that aims to explore the properties of the AMF(k) with respect to filtering, prediction and parameter estimation. There are three groups of simulations, where each group is characterized by the distribution of the state innovation. We examine a Gaussian mixture with two components having both mean zero but different variances; a Gaussian mixture with two components having the same variance but different means; and a Student t distribution with three degrees of freedom.[1]

A first objective of the simulations is an assessment of how close the results of the AMF and the Kalman Filter come to those of the exact filter. Since the number of components of the exact filter increases exponentially with time, such comparisons can only be undertaken for small time series. For time series of length $T = 10$ we explore the discrepancy between the AMF and the Kalman filter on one side and the exact filter on the other side. Such a comparison is carried out with respect to the mean and variance of the filtered and predicted state as well as for the prediction and filtering density at $T = 10$.

Second, for longer time series of length $T = 350$ we compare the performance of the AMF and the Kalman filter in filtering and prediction. Of course, for such long time series the exact filter cannot be employed anymore. We also assess how the relative performance of the two algorithms depends on the parameterization of the data generating state space model. For this type of simulation we treat the hyperparameters of the state space models under consideration as known.

Third, again for time series of length $T = 350$, a subset of the hyperparameters constituting the state space model is treated as unknown. The parameters are estimated using the likelihood methods based on the Kalman

[1] There are similar simulation studies in the literature which try to assess the properties of various collapsing schemes for Gaussian mixtures appearing in state space models. See, e.g., [80], [32], and [50].

filter and the AMF. The distributions of the estimated parameters, obtained from using different algorithms, are compared to each other.

For the group of simulations that uses the Student t distribution, the exact filter algorithm is not known and thus cannot be compared to the AMF and the Kalman filter. However, hyperparameters are estimated even for this case: the state innovation is falsely assumed to be distributed as a normal or a normal mixture and is then estimated using the Kalman filter and the AMF, respectively.

All of the simulations are conducted using GAUSS 3.6. Standard Gaussian pseudo random variables and pseudo random variables from the uniform distribution over $[0, 1]$ are generated using GAUSS's `rndn()` and `rndu()` function respectively. For generating pseudo random variables from a Student t distribution, the function `rstudent()` from the DISTRIB library by Rainer Schlittgen and Thomas Noack is used. The appendix shows how random draws from the Gaussian Mixture are generated.

Numerical maximization of the likelihoods is performed using the BFGS algorithm[2] as implemented in GAUSS's MAXLIK library. The convergence criterion for the gradient is set to $0.5 \cdot 10^{-6}$. Gradients of the likelihood are computed numerically.

7.1 Sampling from a Unimodal Gaussian Mixture

7.1.1 Data Generating Process

For the first block of simulations, data are generated according to a state space model with scalar state process and bivariate measurement vector. We specify the transition equation as

$$\alpha_t = T\alpha_{t-1} + \eta_t \tag{7.1}$$

with

$$\eta_t \sim \omega_1 N(0, Q_1) + (1 - \omega_1) N(0, Q_2). \tag{7.2}$$

The variance of the state innovation is computed as

$$Var(\eta_t) = \omega_1 Q_1 + (1 - \omega_1) Q_2 =: Q. \tag{7.3}$$

The measurement equation is given by

$$\begin{pmatrix} y_{1t} \\ y_{2t} \end{pmatrix} = Z\alpha_t + \epsilon_t \tag{7.4}$$

with

$$\epsilon_t \sim N(0, H). \tag{7.5}$$

[2] See, e.g. [49].

Furthermore, ϵ_t and η_t are each serially independent and independent from each other.

The variance-covariance matrix of the measurement error is assumed to be diagonal with equal variances on the main diagonal,

$$H = h \cdot I_2 =: r \cdot Q \cdot I_2. \tag{7.6}$$

By formulating h as a multiple of Q, the factor r can be interpreted as the inverse signal-noise ratio.

The initial state is generated from a normal with mean and variance being equal to the unconditional mean and variance of the state process, respectively:

$$\alpha_0 \sim N\left(0, \frac{Q}{1-T^2}\right). \tag{7.7}$$

Numerical values are assigned to the system matrices as

$$Z = \begin{pmatrix} 2 \\ -1 \end{pmatrix}, \quad T = 0.5, \quad \omega_1 = 0.15, \quad Q_1 = 1, \quad Q_2 = 50, \quad r = 1, \tag{7.8}$$

which implies a variance of $Q = 8.35$ for the state innovation and a variance of $h = 12.525$ for the measurement error. We will refer to this setting of the parameters as the baseline scenario. The interpretation of this scenario is as follows: with a probability of 85%, innovations come from a Gaussian distribution with moderate variance. With a probability of 15% the innovation is drawn from a Gaussian distribution that permits the occurrence of extreme realizations on both sides of the mean. Figure 7.1 shows typical realizations of the state and the measurement process.[3]

For illustrating the increased probability of obtaining extreme observations compared to the case of a simple normal, we compute the probability of drawing a realization of η_t that is outside the interval $[-3\sqrt{Q}, +3\sqrt{Q}]$. When η_t has the specified mixture distribution we obtain[4]

$$\begin{aligned} &1 - P(-3\sqrt{Q} < \eta_t < 3\sqrt{Q}) \\ &= 2\left(\omega_1 \Phi(-3\frac{\sqrt{Q}}{\sqrt{Q_1}}) + (1-\omega_1)\Phi(-3\frac{\sqrt{Q}}{\sqrt{Q_2}})\right) = 0.033, \end{aligned}$$

whereas for the case of a simple normal with the same variance we get

$$1 - P(-3\sqrt{Q} < \eta_t < 3\sqrt{Q}) = 2\Phi(-3) = 0.0027,$$

which is more then ten times smaller.

[3] Admittedly, the qualifier 'typical' has a strong flavor of subjectiveness. Here and in the following, we will call an outcome of a random experiment 'typical', if the outcome exhibits certain patterns that arise in the majority of outcomes of the same experiment.

[4] Recall that Φ denotes the cumulative distribution function of the standard normal distribution.

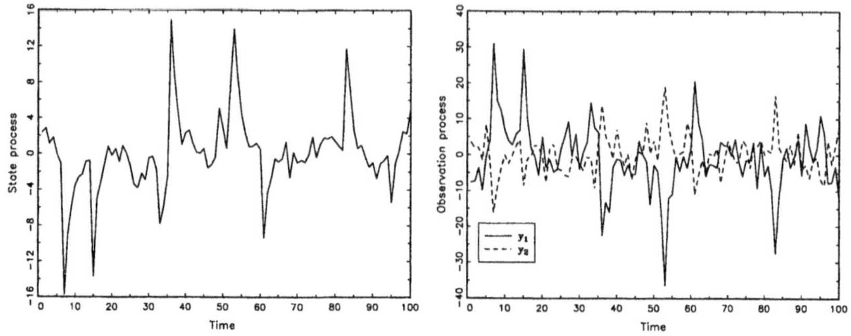

Fig. 7.1. Typical realization of the state and observation process, DGP with unimodal innovation density, $T = 100$ observations

7.1.2 Filtering and Prediction for Short Time Series

For a typical realization $\{y_t\}$, $t = 1, \ldots, 10$, of the observation process, the left panel of figure 7.2 contains sequences $\{a_{t|t}\}$ of filtered states obtained from using the Kalman filter, the AMF(1) and the AMF(3). Here and in all other estimations considered in this section, the filtering algorithm is initialized with

$$a_0 = 0, \quad \Sigma_0 = \frac{Q}{1 - T^2}.$$

The results corresponding to the AMF(1), the AMF(3) and the exact filter are indistinguishable. The Kalman filter departs from the exact filter substantially but tracks its changes in direction quite well. Results of the same nature hold for the predicted state process $\{a_{t|t-1}\}$, $t = 1, \ldots, 10$, in the right panel of figure 7.2.

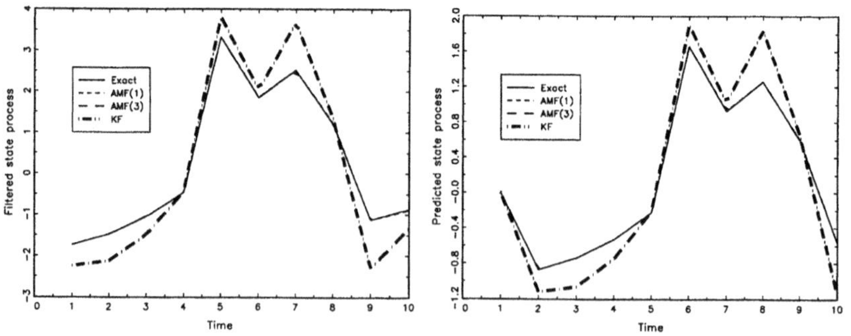

Fig. 7.2. Filtered and predicted state process, DGP with unimodal innovation density, $T = 10$

In figure 7.3 a similar comparison is made for the process of the conditional variances of the state. The exact filter yields the sequences $\{\Sigma_{t|t}\}$ and $\{\Sigma_{t|t-1}\}$ of exact conditional variances of the state given the observations. That is $\Sigma_{t|t} = Var(\alpha_t|\mathcal{Y}_t)$ and $\Sigma_{t|t} = Var(\alpha_t|\mathcal{Y}_{t-1})$ for $t = 1, \ldots, T$. The approximated variances generated by the AMF(1) differ only slightly from their exact counterparts whereas the variances generated by the AMF(3) coincide with the exact solution. As discussed above in chapter 5, the sequences of $\{\Sigma_{t|t}\}$ and $\{\Sigma_{t|t-1}\}$ generated by the Kalman filter correspond to the MSEs of the state estimators. They do not have the interpretation of conditional variances and thus can not be fully compared to the other lines in the picture.

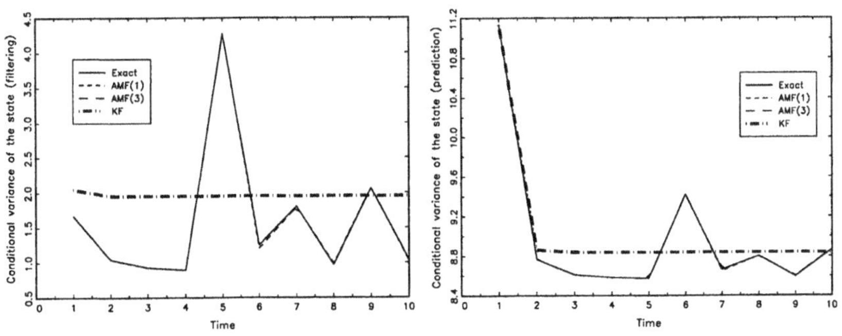

Fig. 7.3. Conditional variances of the state, filtering and prediction, DGP with unimodal innovation density, $T = 10$

For a small t it is also possible to visualize the true conditional densities $p(\alpha_t|\mathcal{Y}_t)$ and $p(\alpha_t|\mathcal{Y}_{t-1})$ of the state. Figure 7.4 presents the exact filtering density $p(\alpha_{10}|\mathcal{Y}_{10})$ (left panel) and the exact prediction density $p(\alpha_{10}|\mathcal{Y}_9)$ (right panel) at $T = 10$ as a solid line. The densities correspond to the same 10 observations that have been used to generate figures 7.2 and 7.3. Both the filtering and the prediction densities are mixtures of 1024 single normal densities. The dashed-dotted lines[5] in the panels depict the normal densities implied by the Kalman filter, i.e. $\phi(\alpha_{10}; a^{KF}_{10|10}, \Sigma^{KF}_{10|10})$ and $\phi(\alpha_{10}; a^{KF}_{10|9}, \Sigma^{KF}_{10|9})$, respectively. The short-dashed and long-dashed lines represent the mixture distributions generated from the AMF(1) and the AMF(3) respectively. The AMF(1) density, a Gaussian mixture with two components, comes very close to the exact conditional density, the AMF(3) density, a Gaussian mixture with eight components is indistinguishable from it. Figure 7.5 contains the same kind of densities as figure 7.4, but for a different realization of the observation process. It serves to document that the filtering density can in fact be bimodal. This is quite interesting since all distributions appearing in the data generating process are unimodal.

[5] They are also slightly bolder than the other lines in the picture.

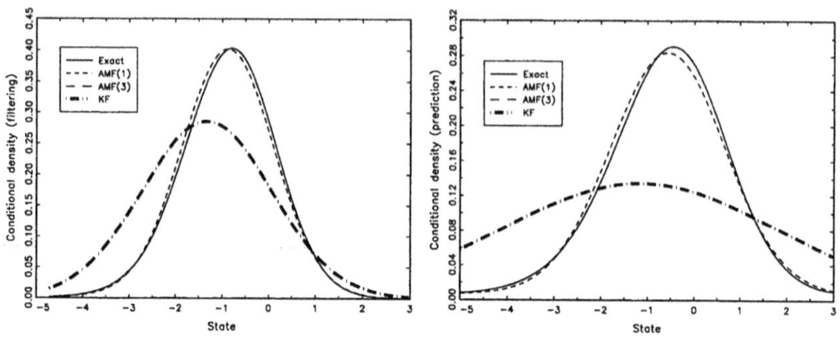

Fig. 7.4. Conditional density of the state at $T = 10$, DGP with unimodal innovation density, filtering and prediction: a unimodal example

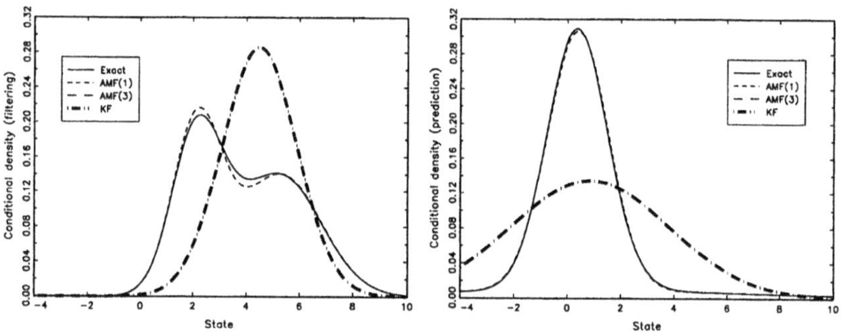

Fig. 7.5. Conditional density of the state at $T = 10$, DGP with unimodal innovation density, filtering and prediction: a bimodal example

In addition to the visualization of the typical outcomes using the figures above, we quantify the discrepancy between the results of the exact filter, the AMF(1), the AMF(3) and the Kalman filter by the following simulation. Using parameterization (7.8), 1000 realizations of the observation process $\{y_t\}$ of length $T = 10$ are generated. At the final observation at $T = 10$, the exact filtering density and the exact prediction density are both mixtures of 2^{10} normal densities. The mean and variance components as well as the weights of these mixtures are generated by the exact filter. Let $a^{ex}_{10|10}$ and $a^{ex}_{10|9}$ denote the corresponding means and $\Sigma^{ex}_{10|10}$ and $\Sigma^{ex}_{10|9}$ denote the corresponding variances of these mixtures. For each run of the simulation we compute the absolute deviation between these four quantities and the corresponding expressions generated by the AMF(1), the AMF(3) and the Kalman filter. The absolute deviations are averaged over the 1000 runs. The results are given in table 7.1. With respect to the here selected measure of deviation, the results of both versions of the AMF are very close to the exact filter outcomes while the

Kalman filter output shows a substantial discrepancy. Moreover, it turns out that the AMF(3) is closer to the exact filter than the AMF(1), especially with respect to the variance of the filtering density.

Table 7.1. Deviations of KF, AMF(1) and AMF(3) from the exact filter, DGP with unimodal innovation density

Deviation for algorithm f	KF	AMF(1)	AMF(3)
$\overline{\lvert a^{f}_{10\mid 10} - a^{ex}_{10\mid 10}\rvert}$	0.648	6.5e-3	4.5e-4
$\overline{\lvert \Sigma^{f}_{10\mid 10} - \Sigma^{ex}_{10\mid 10}\rvert}$	0.891	1.3e-2	3.2e-4
$\overline{\lvert a^{f}_{10\mid 9} - a^{ex}_{10\mid 9}\rvert}$	0.330	3.0e-3	7.9e-4
$\overline{\lvert \Sigma^{f}_{10\mid 9} - \Sigma^{ex}_{10\mid 9}\rvert}$	0.223	3.1e-3	4.1e-4

Mean absolute deviations between the exact conditional expectation and variance and the corresponding quantities generated by the KF, the AMF(1) and the AMF(3) at $T = 10$.

7.1.3 Filtering and Prediction for Longer Time Series

For short time series we have been able to compare the performance of the exact filter to the performance of the AMF and the Kalman filter. For longer time series, computation of the exact filter is no longer feasible. We therefore focus on the relative performance of the AMF compared to the optimal linear filter. To this end simulations of the following type are conducted.

For each run a bivariate time series $\{y_t\}$ of length $T' = 370$ is generated according to the data generating process specified by (7.1) - (7.7). The first 20 observations are deleted. This serves to prevent any dependence on initial conditions. To the resulting time series of length $T = 350$, the AMF(1) and the Kalman filter are applied yielding series $\{a_{t|t}\}$ and $\{a_{t|t-1}\}$, $t = 1, \ldots, 350$, of filtered and predicted states. Based on these sequences mean squared errors between the true state and the filtered or predicted state are computed ignoring the first 10 elements of the respective series. That is we compute for the ith run of the simulation:

$$MSE^{fi}_{f,i} = \frac{1}{340} \sum_{t=11}^{350} \left(\alpha_t - a^{f}_{t|t}\right)^2, \quad f = KF,\, AMF(1)$$

and

$$MSE^{pr}_{f,i} = \frac{1}{340} \sum_{t=11}^{350} \left(\alpha_t - a^{f}_{t|t-1}\right)^2, \quad f = KF,\, AMF(1).$$

The ratio γ of the two MSEs,

$$\gamma_i^{fi} = \frac{MSE_{KF,\,i}^{fi}}{MSE_{AMF,\,i}^{fi}}, \quad \gamma_i^{pr} = \frac{MSE_{KF,\,i}^{pr}}{MSE_{AMF,\,i}^{pr}},$$

is computed for each run and will be used as measure of relative efficiency.[6] We will refer to $\gamma - 1$ as the efficiency gain obtained from using the AMF instead of the Kalman filter.

For filtering and prediction we only compare the Kalman filter and the AMF(1) leaving results from the AMF algorithm with $k > 1$ aside. This is done because the results from the AMF(1) and the AMF(3) turn out to be nearly identical for the scenarios considered. In light of the previous subsection this can be interpreted as desirable: already for $k = 1$, the AMF output is nearly equal to the output of the exact filter, and an increase in k does not lead to any substantial change of the results.

The simulation is carried out for different parameter settings. Starting from the baseline scenario (7.8) we vary either the mean reversion parameter T of the state process, the weight ω_1 in the mixture distribution of the state innovation, the variance Q_1 in the mixture, or the inverse signal-noise ratio r. We only change one of these parameters at a time, holding all other parameters fixed on their baseline values. Each of the parameters takes on four different values apart from the baseline value, thus we perform 1+4+4+4+4=17 simulations. Each simulation uses 1000 runs.

Table 7.2 summarizes the results. The right column shows the parameter setting. $T = 0.3$, e.g., stands for the case where $T = 0.3$ and all other parameters are held fixed on their baseline values. The columns 'Mean' and 'Std Dev' contain the means and the standard deviations of the relative efficiency measure γ over the 1000 runs for each scenario. The quantity 'Frac' is the fraction of runs for which $\gamma > 1$, i.e. for which the MSE of the AMF(1) has been smaller than that of the Kalman filter.

The distribution of relative efficiency γ for the different scenarios is characterized by groups of boxplots in figures 7.6 (filtering) and 7.7 (prediction).[7] Note that the boxplot corresponding to the baseline scenario is contained in all of the panels.

As expected, relative efficiency is equal to one for all runs if $Q_1 = 1$. In this case, the distribution of the innovation density is a mixture of two standard normals, and the AMF(1) coincides with the Kalman filter. In all other

[6] Such a measure is also employed in the simulations conducted by [51].

[7] Here and in the following, for a data set $x_1, \ldots, x_n$ the boxplot's lower whisker corresponds to $\max\{x_{(1)}; x_{0.25} - 1.5IQR\}$, and the upper whisker corresponds to $\min\{x_{(n)}; x_{0.75} + 1.5IQR\}$. $x_{(1)}$ is the smallest observation, $x_{(n)}$ is the largest observation, $x_{0.25}$ is the 25%-quantile, $x_{0.75}$ is the 75%-quantile, and $IQR = x_{0.75} - x_{0.25}$ is the interquartile range. This type of boxplot is described under the name modified boxplot ("modifierter Boxplot") by [48]. It differs from the standard offered by GAUSS.

Table 7.2. Relative efficiency of AMF(1) vs. KF for different parameter settings, DGP with unimodal innovation density

	Filtering			Prediction			
#	Mean	Std Dev	Frac	Mean	Std Dev	Frac	Variant
1	1.477	0.123	1.000	1.018	0.016	0.879	baseline
2	1.343	0.112	1.000	1.015	0.014	0.881	$\omega_1 = 0.05$
3	1.354	0.098	1.000	1.014	0.013	0.857	$\omega_1 = 0.25$
4	1.239	0.071	1.000	1.011	0.012	0.827	$\omega_1 = 0.35$
5	1.148	0.050	0.999	1.007	0.010	0.776	$\omega_1 = 0.45$
6	1.000	-	0.000	1.000	-	0.000	$Q_1 = 1$
7	1.088	0.041	0.991	1.004	0.008	0.729	$Q_1 = 10$
8	1.718	0.181	1.000	1.024	0.017	0.916	$Q_1 = 100$
9	1.966	0.236	1.000	1.028	0.019	0.932	$Q_1 = 200$
10	1.580	0.139	1.000	1.001	0.003	0.612	$T = 0.1$
11	1.545	0.136	1.000	1.007	0.010	0.788	$T = 0.3$
12	1.381	0.107	1.000	1.030	0.020	0.939	$T = 0.7$
13	1.275	0.082	1.000	1.041	0.023	0.970	$T = 0.95$
14	1.315	0.074	1.000	1.006	0.008	0.757	$r = 0.5$
15	1.429	0.103	1.000	1.012	0.012	0.841	$r = 1.0$
16	1.482	0.133	1.000	1.023	0.017	0.913	$r = 2.0$
17	1.467	0.140	1.000	1.026	0.018	0.972	$r = 2.5$

scenarios considered, mean relative efficiency is bigger than 1. The following comments apply to the cases with $Q_1 \neq 1$ only.

Concerning the filtering problem, the fraction of runs in which the AMF(1) has a smaller MSE than the Kalman filter is 100% in most cases. The smallest fraction of 99.1% shows up for the case in which the variance ratio of the mixture components is only 10. The mean efficiency gain ranges from 8.8% (scenario $Q_1 = 10$ again) to 96.6% (scenario $Q_1 = 100$).

For the mean reversion parameter T of the state process and the variance ratio Q_1, a quite unambiguous relationship emerges between the parameter values and the gain in efficiency: average relative efficiency $\bar{\gamma}^{fi}$ decreases in T and rises in Q_1. The latter result may be characterized as intuitive, since the bigger the value of Q_1 the more the nature of the innovation density departs from that of a simple normal.

The dependence of relative efficiency on ω_1 and r is a little less clear cut. The lower right panel of figure 7.6 suggests that the mean efficiency gain remains quite constant at about 47% from $r = 1.5$ to $r = 2.5$. For the five values of ω_1 considered here, mean relative efficiency has a maximum at $\omega_1 = 0.15$. It is quite intuitive that $\bar{\gamma}_{fi}$ viewed as a function of ω_1 must have at least one local maximum on $(0, 1)$. Moving towards the edges of that interval corresponds to moving towards a simple normal distribution of the state innovation which in turn implies a relative efficiency of one, while values of ω_1 within the interval correspond to a true mixture.

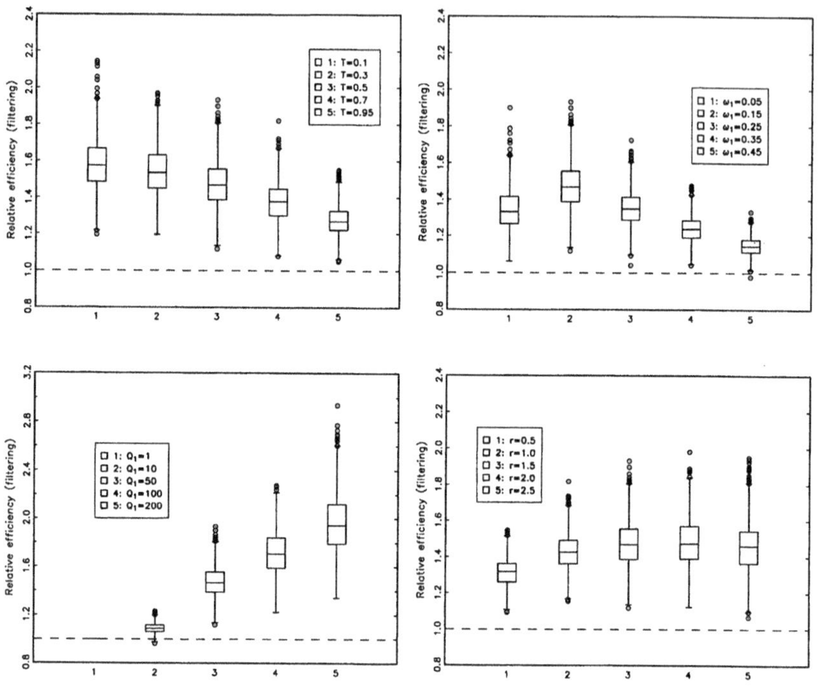

Fig. 7.6. Relative efficiency of filtering for different parameter settings, DGP with unimodal innovation density

Turning now to the results concerning prediction, figure 7.7 basically shows the same shapes as figure 7.6 with two main exceptions. First, relative efficiency now increases as a function of T. Second, the efficiency gain from using the AMF(1) as opposed to the Kalman filter is consistently lower than for the filtering problem.

The latter result can be understood heuristically by recalling the relationship between the filtering MSE and the prediction MSE. For the Kalman filter as well as for the AMF we have

$$MSE(\alpha_{t+1}, a_{t+1|t}) = T^2 MSE(\alpha_t, a_t) + Q$$

for a scalar state process. Let us assume for a moment that both the filtering and the prediction MSE do not depend on time for either of the two filters.[8] We have then that

$$\gamma^{fi} = \frac{MSE_{KF}^{fi}}{MSE_{AMF}^{fi}}, \quad \text{and} \quad \gamma^{pr} = \frac{T^2 MSE_{KF}^{fi} + Q}{T^2 MSE_{AMF}^{fi} + Q}. \tag{7.9}$$

[8] We know that this is true for the Kalman filter when $\Sigma_{t|t}$ and $\Sigma_{t|t-1}$ have converged to their steady state values.

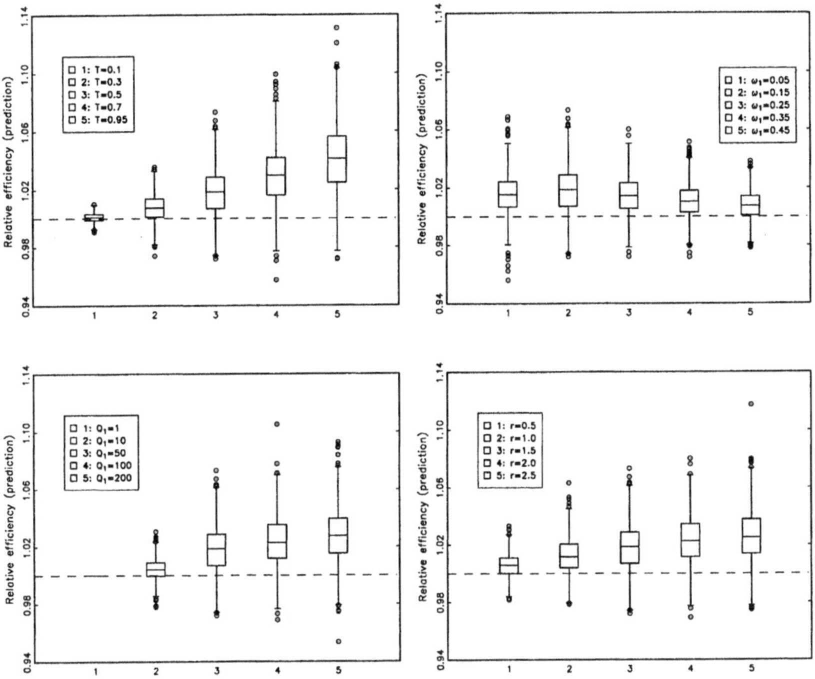

Fig. 7.7. Relative efficiency of prediction for different parameter settings, DGP with unimodal innovation density

If the AMF beats the Kalman filter with respect to filtering, i.e. $\gamma^{fi} > 1$, then (7.9) implies that $\gamma^{pr} < \gamma^{fi}$. This can be shown by computing the ratio of γ^{fi} and γ^{pr}:

$$\begin{aligned}
\frac{\gamma^{pr}}{\gamma^{fi}} &= \frac{T^2 MSE^{fi}_{KF} + Q}{T^2 MSE^{fi}_{AMF} + Q} \cdot \frac{MSE^{fi}_{AMF}}{MSE^{fi}_{KF}} \\
&= \frac{T^2 MSE^{fi}_{AMF} + Q \frac{MSE^{fi}_{AMF}}{MSE^{fi}_{KF}}}{T^2 MSE^{fi}_{AMF} + Q} \\
&< \frac{T^2 MSE^{fi}_{AMF} + Q}{T^2 MSE^{fi}_{AMF} + Q} \\
&= 1,
\end{aligned}$$

where the inequality follows from $MSE^{fi}_{KF}/MSE^{fi}_{AMF} = \gamma^{fi} > 1$.

So we have seen that relative efficiency is generally lower for prediction than for filtering. However, mean relative efficiency for prediction is still greater than one for all scenarios considered. Using that $\gamma^{fi} > 1$, this simulation result can again be deduced from (7.9):

$$\gamma^{pr} = \frac{T^2 MSE_{KF}^{fi} + Q}{T^2 MSE_{AMF}^{fi} + Q} > \frac{T^2 MSE_{AMF}^{fi} + Q}{T^2 MSE_{AMF}^{fi} + Q} = 1.$$

The mean efficiency gain ranges from 0.1% (scenario $T = 0.1$) to 4.1% (scenario $T = 0.95$). The fraction of runs in which the AMF(1) has been superior to the Kalman filter goes from 61.2% (scenario $T = 0.1$) to 97.2% (scenario $r = 2.5$).

All of these simulation results can only convey a first impression of how a change in parameters affects the efficiency gain of the AMF(1) over the optimal linear filter. If one is interested in the dependence of relative efficiency on a specific parameter (or set of parameters) one would conduct similar simulations as those described, but for a finer grid for the parameter of interest.

7.1.4 Estimation of Hyperparameters

Up to now the focus has been only on the filtering and prediction problem, treating the model parameters as known. Most of the simulation studies in the literature that explore estimation techniques for mixture state space models are of this kind. For the application of mixture state space models in fields like economics, however, it is of considerable interest to learn about the properties of the different algorithms with respect to parameter estimation. The following simulation explores how well unknown parameters can be estimated. Estimation is conducted using both the quasi-likelihood based on the Kalman filter and the approximate likelihood based on the AMF(1). We already remark here that the parameter estimates based on the AMF(3) have been so similar to those from the AMF(1) that only the latter are reported.

The simulation is based on 1000 runs. For each run we simulate time series of 370 observations according to the baseline scenario (7.8). We then drop the first 20 observations and use the remaining 350 observations for estimation. The model parameters Q_1, Q_2, ω_1, T and h are treated as unknown.[9] We estimate these parameters using the AMF(1).

Using the quasi-likelihood based on the Kalman filter it is not possible to separately estimate the variance components Q_1, Q_2, and ω_1. The quasi-likelihood is a function of Q as a whole only, implying that there are infinitely many triples (Q_1, Q_2, ω_1) that imply the same value of Q and thus the same value of the approximate likelihood function. Consequently, we apply the Kalman filter to estimate T, h and Q.

The filters are again initialized with

$$a_0 = 0, \quad \Sigma_0 = \frac{Q}{1 - T^2}$$

[9] Recall that h is the diagonal element of the variance-covariance matrix of the measurement error, i.e. $H = h \cdot I_2$.

where now of course T and Q depend on the unknown parameters.

The approximate likelihoods are maximized numerically using the BFGS algorithm.[10] In order to stabilize the optimization algorithm, the unknown model parameters are reparameterized in a vector ψ. The parameterization is chosen in such a way that for the true values of the model parameters, the components of ψ have nearly the same dimension. Moreover, it is made sure that certain restrictions on the model parameters are satisfied.

For the estimation based on the AMF(1) we set

$$T = \frac{\psi_1}{10}, \quad \omega_1 = \left(\frac{\psi_2}{10}\right)^2, \quad Q_1 = \psi_3^2, \quad Q_2 = \left(\frac{\psi_4}{5}\right)^2, \quad h = \psi_5^2, \tag{7.10}$$

which implies positivity restrictions on all variance components and on ω_1. The vector ψ that corresponds to the numerical values of the baseline scenario is given as

$$\psi^* = (5.0,\ 3.873,\ 7.071,\ 5.0,\ 3.539)'.$$

As starting values for the optimization of the approximate likelihood functions we choose

$$\psi^0 = (5.0,\ 4.0,\ 7.0,\ 5.0,\ 4.0)'$$

for all runs.[11]

For the estimation based on the Kalman filter the parameterization is chosen as

$$T = \frac{\psi_1}{10}, \quad Q = \psi_2^2, \quad h = \psi_3^2. \tag{7.11}$$

The true value of ψ is

$$\psi^* = (5.0,\ 2.890,\ 3.539)'.$$

As starting values for all runs we choose

$$\psi^0 = (5.0,\ 3.0,\ 4.0)'.$$

After each run, the components of the estimate $\hat{\psi}$ are retransformed using (7.10) and (7.11), respectively, to obtain estimates of the model parameters themselves. In addition, for the AMF(1), an implied estimate of Q is constructed from the estimates of Q_1, Q_2 and ω_1 as

$$\hat{Q} = \hat{\omega}_1 \hat{Q}_1 + (1 - \hat{\omega}_1)\hat{Q}_2.$$

The parameterizations do not impose a bound on ω_1 from above nor do they constrain the mean reversion parameter T to stationarity. Without imposing these constraints, however, the estimates of T are all within the interval $[0.3, 0.7]$ and all estimated ω_1 are below 0.35.

---

[10] See the remarks on page 102.

[11] Alternatively, one could draw starting values at random from some interval which has also been tried. For all of the simulations, maximizers did not depend on starting values.

Table 7.3 summarizes the estimation results. For an interpretation of the results it has to be kept in mind that the AMF estimates five parameters whereas the Kalman filter only estimates three. The relative bias for the estimates of T, Q and h is smaller than 0.5% for both algorithms.

The standard deviation is slightly smaller for the AMF(1), especially in case of the mean reversion parameter T. The distribution of these estimates is characterized by means of boxplots in the left three panels of figure 7.8. The distributions of $\hat{T}$ and $\hat{h}$ look quite symmetric, while the distribution of $\hat{Q}$ is a little skewed to the right.

The rightmost panel of figure 7.8 describes the distribution of relative efficiency (for filtering) over the thousand runs. The average ratio of the MSEs for the 1000 runs has been $\bar{\gamma}^{fi} = 1.447$. The efficiency gain has been positive for all runs of the simulation. Moreover, the efficiency gain is quite similar to that obtained for the baseline scenario with known parameters.[12]

The estimates of the individual components of the innovation density exhibit a stronger bias: 4%, 0.86% and 5.4% for $\hat{\omega}_1$, $\hat{Q}_1$ and $\hat{Q}_2$ respectively. The distributions of all of theses estimators are skewed to the right as shown in figure 7.9. It is quite interesting to observe that although the estimated single components of the mixture variance show standard deviation and biases which are quite large, the implied estimator of the overall variance Q has moderate bias (0.49% only) and a relatively smaller standard deviation.

Table 7.3. ML estimates of parameters, DGP with unimodal innovation density

		Mean of estimate		Standard deviation	
Parameter	true	KF	AMF(1)	KF	AMF(1)
T	0.500	0.499	0.500	0.056	0.038
Q	8.350	8.309	8.318	1.811	1.805
h	12.525	12.503	12.513	0.958	0.923
ω_1	0.15	-	0.156	-	0.040
Q_1	50.000	-	50.431	-	14.133
Q_2	1.000	-	0.946	-	0.441

The advantage of the AMF(1) over the Kalman filter is the possibility to estimate the individual components of the mixture distribution of the state innovation. Thus, each set of estimates $\hat{\omega}_1$, $\hat{Q}_1$ and $\hat{Q}_2$, obtained from the AMF implies an estimated density for the state innovation η_t. The Kalman filter on the other hand implicitly assumes a Gaussian η_t and estimates only the variance of the state innovation as a whole. We close this section with a graphical comparison of the true innovation density with the 'average' estimated density. Such a comparison can be made in a variety of ways. For example one could compare the mixture corresponding to the true values with the one corresponding to the estimates $\bar{\hat{\omega}}_1$, $\bar{\hat{Q}}_1$ and $\bar{\hat{Q}}_2$ averaged over the 1000 runs.

[12] See figure 7.6 and table 7.2 above.

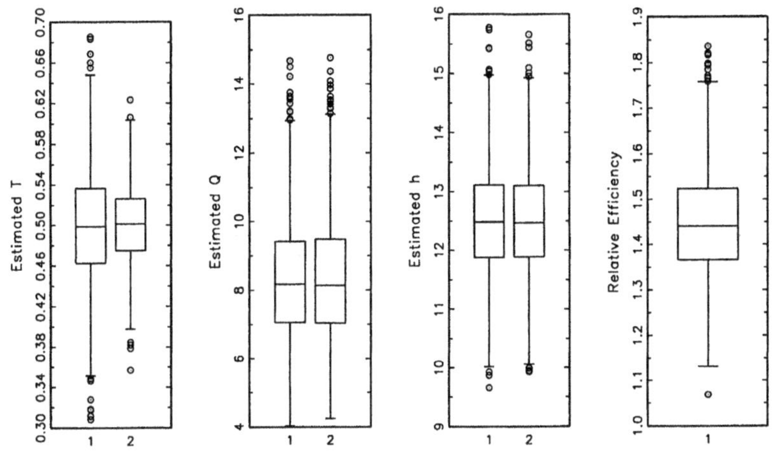

Fig. 7.8. Distribution of parameter estimates (1) and relative efficiency, DGP with unimodal innovation density. For the panels with two boxplots, the left one corresponds to the Kalman filter, the right one corresponds to the AMF(1).

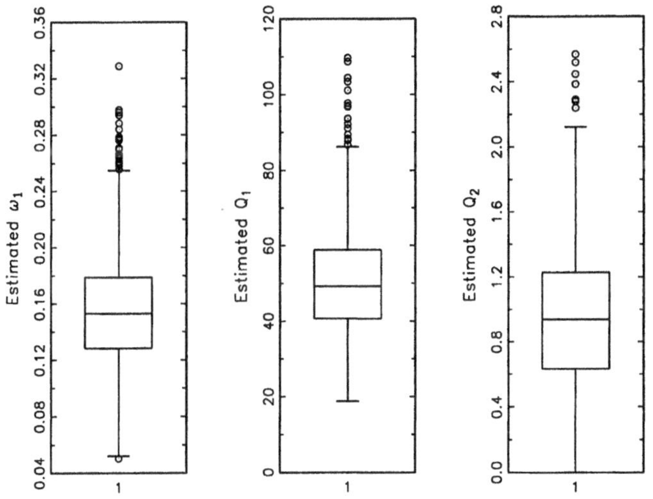

Fig. 7.9. Distribution of parameter estimates (2), DGP with unimodal innovation density

In the comparison made here, we average directly over the densities which correspond to the different parameter values. Let $\hat{Q}_1^i$, $\hat{Q}_2^i$ and $\hat{\omega}_1^i$ denote the model parameters estimated for the ith run. The short-dashed line in figure 7.10 corresponds to the averaged estimated density[13]

[13] It is easily observed that $\bar{p}_{AMF}(\eta_t)$ is indeed a valid probability density function. – A density curve in figure 7.10 is constructed by using a fine grid of points within

$$\bar{\hat{p}}_{AMF}(\eta_t) = \frac{1}{1000} \sum_{i=1}^{1000} \left(\hat{\omega}_1^i \, \phi(\eta_t;\, 0,\, \hat{Q}_1^i) + (1 - \hat{\omega}_1^i) \, \phi(\eta_t;\, 0,\, \hat{Q}_2^i) \right) \tag{7.12}$$

while the solid line is the true model density

$$p(\eta_t) = 0.15 \, \phi(\eta_t;\, 0,\, 50) + 0.85 \, \phi(\eta_t;\, 0,\, 1). \tag{7.13}$$

The long-dashed line in the figure is the average of the simple normal densities implied by the parameter estimates based on the Kalman filter,

$$\bar{\hat{p}}_{KF}(\eta_t) = \frac{1}{1000} \sum_{i=1}^{1000} \phi(\eta_t;\, 0,\, \hat{Q}^i).$$

The shape of the average probability density function (pdf) implied by the AMF(1) matches the shape of the true pdf quite well, but it exhibits some excess probability mass in the center. The average pdf implied by the Kalman filter cannot replicate the shape of the true pdf.

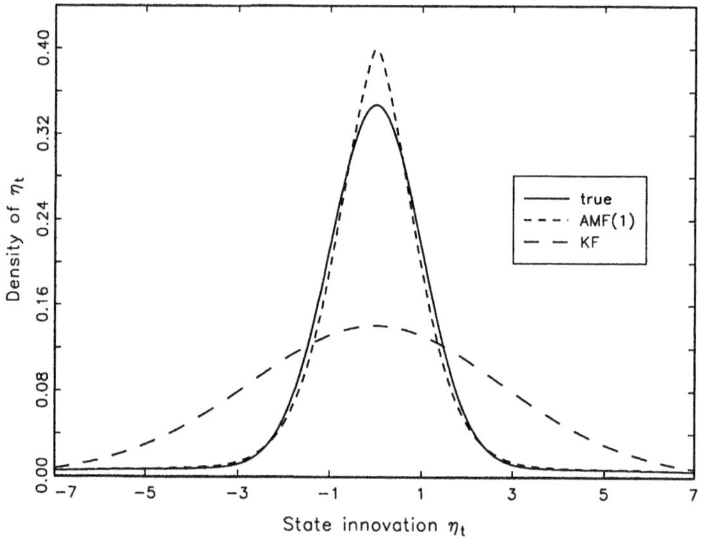

Fig. 7.10. True and estimated 'average' density of the state innovation, DGP with unimodal innovation density

---

the interval [-7,7]. For each point $\bar{\eta}_t$ of this grid, the 1000 densities evaluated at that particular point $\bar{\eta}_t$ are averaged.

7.2 Sampling from a Bimodal Gaussian Mixture

7.2.1 Data Generating Process

For this second block of simulations we assume a Gaussian mixture distribution for the state innovation whose components have the same variance but different means. The structure of this subsection is similar to that of the previous one, the same types of simulations are conducted. More comparisons are made between the AMF(1) and the AMF(3), since this time the differences between these two versions of the AMF show up more clearly. An extensive comparison of relative efficiency for different parameter settings is not considered here, however.

Apart from the different specification of the innovation density, the structure of the model is the same as in section 7.1. We write down the whole model again for convenience:

The transition equation is given by

$$\alpha_t = T\alpha_{t-1} + \eta_t, \tag{7.14}$$

with

$$\eta_t \sim \omega_1 N(\mu_1, Q_1) + (1-\omega_1)N(\mu_2, Q_1), \tag{7.15}$$

where

$$\omega_1\mu_1 + (1-\omega_1)\mu_2 = 0. \tag{7.16}$$

For the variance of the state innovation one obtains

$$Var(\eta_t) = \omega_1(Q_1 + \mu_1^2) + (1-\omega_1)(Q_1 + \mu_2^2) =: Q. \tag{7.17}$$

For the measurement equation we have

$$\begin{pmatrix} y_{1t} \\ y_{2t} \end{pmatrix} = Z\alpha_t + \epsilon_t \tag{7.18}$$

with

$$\epsilon_t \sim N(0, H). \tag{7.19}$$

Again, H is a diagonal matrix with equal entries on the main diagonal. The diagonal elements are reparameterized as a multiple of Q,

$$H = h \cdot I_2 =: r \cdot Q \cdot I_2. \tag{7.20}$$

ϵ_t and η_t are each serially independent and independent of each other.

A draw from a simple normal generates the initial state:

$$\alpha_0 \sim N\left(0, \frac{Q}{1-T^2}\right). \tag{7.21}$$

The baseline scenario for this section is given by:

$$T = 0.9, \quad \omega_1 = 0.10, \quad \mu_1 = 9.0, \quad \mu_2 = -1.0, \quad Q_1 = 1.0$$
$$Z = \begin{pmatrix} 2 \\ -1 \end{pmatrix}, \quad r = 2.0. \tag{7.22}$$

The here chosen innovation density is bimodal.[14] It implies that 90% of the time, realizations of η_t are generated from a normal distribution with unit variance which is centered around -1. With a probability of 10%, a realization comes from a normal distribution that has again a variance of one but its location ($\mu_2 = 9$) is far off on the other side of the overall mean (zero) of the distribution. The state process considered in the previous subsection generated extreme observations symmetrically around zero. Here, an extreme observation, if it appears, shows up consistently on the right side of the mean. A typical realization of the state and observation process is shown in figure 7.11.

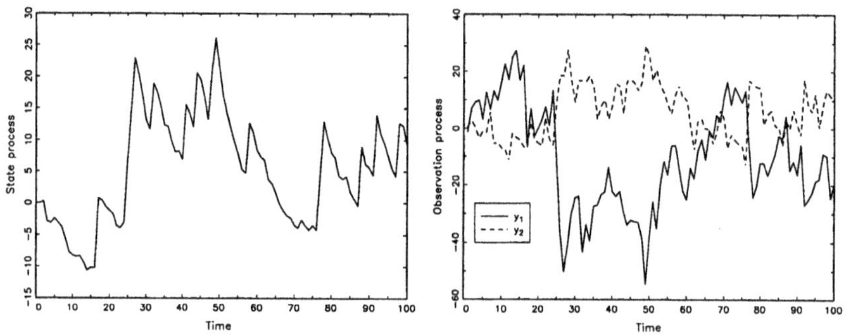

Fig. 7.11. Typical realization of the state and observation process, DGP with bimodal innovation density, $T = 100$ observations

7.2.2 Filtering and Prediction for Short Time Series

Again, for small time series we assess how close the outputs of the AMF(1), the AMF(3), and the Kalman filter come to the output of the exact filter. Figure 7.12 shows a typical situation concerning the results of filtering and prediction. Similar to the results of the previous subsection, the Kalman filter tracks the course of the exactly filtered and predicted state process quite well, but shows substantial differences in level. The AMF results are very close to those from the exact filter.

The sequence of exact conditional variances of the state (solid line) in figure 7.13 is matched well by the approximate variances generated by the

[14] A picture of the density is given as a solid line in figure 7.19 below, in which we will again compare the true and the estimated density.

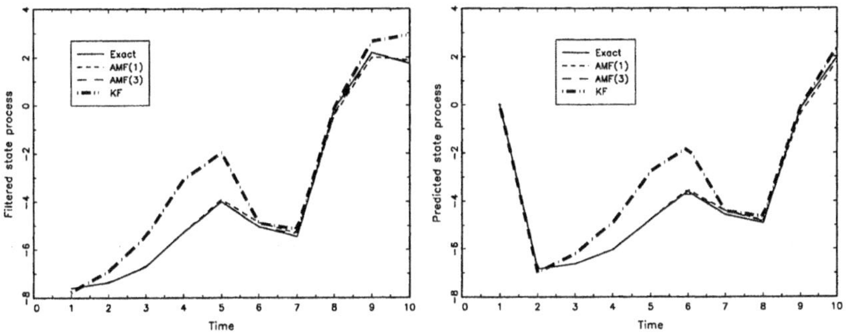

Fig. 7.12. Filtered and predicted state process, DGP with bimodal innovation density, $T = 10$ observations

AMF(1) and the AMF(3). Note, however, that unlike in subsection 7.1, the AMF(3) is distinctly closer to the exact solution than the AMF(1). In fact, in the picture the AMF(3) coincides with the exact solution, whereas for the AMF(1) some discrepancies are visible.

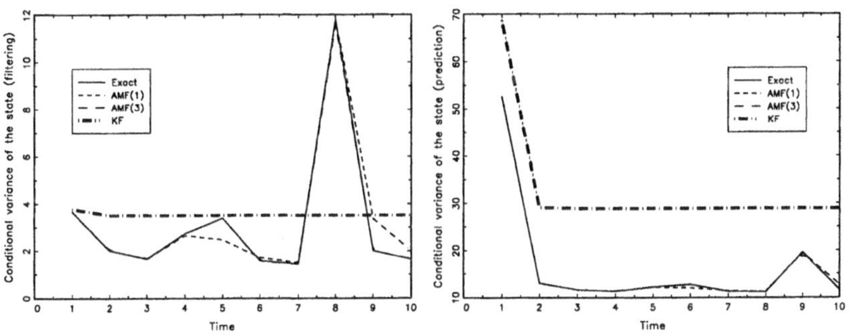

Fig. 7.13. Conditional variances of the state, filtering and prediction, DGP with bimodal innovation density, $T = 10$ observations

For one arbitrary realization of the observation process, figure 7.14 shows the exact conditional densities $p(\alpha_{10}|\mathcal{Y}_{10})$ and $p(\alpha_{10}|\mathcal{Y}_9)$ together with their estimated counterparts. For both the filtering and the prediction case, the AMF(3) exhibits a considerably better matching of the exact densities than the AMF(1).

Finally, we again compute the average absolute distance between the outputs of the exact filter and the various approximations using 1000 simulated time series of length 10. There is a clear rank order with respect to accuracy.

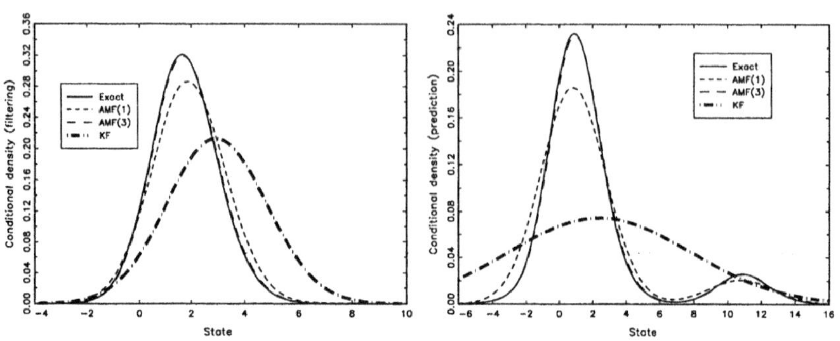

Fig. 7.14. Conditional density of the state at $T = 10$, DGP with bimodal innovation density, filtering and prediction

The AMF(3) ranks before the AMF(1) which in turn ranks before the Kalman filter.

Table 7.4. Deviations of KF, AMF(1) and AMF(3) from the exact filter, DGP with bimodal innovation density

Type of algorithm f	KF	AMF(1)	AMF(3)
$\overline{\lvert a^f_{10\vert 10} - a^{ex}_{10\vert 10}\rvert}$	0.952	0.069	0.020
$\overline{\lvert \Sigma^f_{10\vert 10} - \Sigma^{ex}_{10\vert 10}\rvert}$	2.045	0.119	0.021
$\overline{\lvert a^f_{10\vert 9} - a^{ex}_{10\vert 9}\rvert}$	0.860	0.064	0.030
$\overline{\lvert \Sigma^f_{10\vert 9} - \Sigma^{ex}_{10\vert 9}\rvert}$	17.343	0.103	0.040

Mean absolute deviations between the exact conditional expectation and variance and the corresponding quantities generated by the KF, the AMF(1) and the AMF(3) at $T = 10$.

7.2.3 Filtering and Prediction for Longer Time Series

In the same fashion as in section 7.1.3, 1000 time series of length $T = 350$ are generated and the filtering and prediction performance of the different algorithms is compared. Besides results from the AMF(1) and the Kalman filter, we also include results from the AMF(3). Three simulations are conducted: one is based on parameter values from the baseline setting (7.22), the other two set $Q_1 = 0.5$ and $Q_1 = 0.05$, respectively, while leaving all other parameter values on the baseline values. Smaller values of the variances 'sharpen' the bimodal structure of the innovation density: the smaller the value of Q_1 , the more probability mass is concentrated around the two centers μ_1 and μ_2.

Figure 7.15 and table 7.5 document the efficiency gain of the AMF(1) over the Kalman filter. For filtering as well as for prediction the improvement in precision decreases in Q_1. The mean efficiency gain is large, reaching from 10.2% for the prediction problem with $Q_1 = 1.0$ to 584% for the filtering problem with $Q_1 = 0.05$.

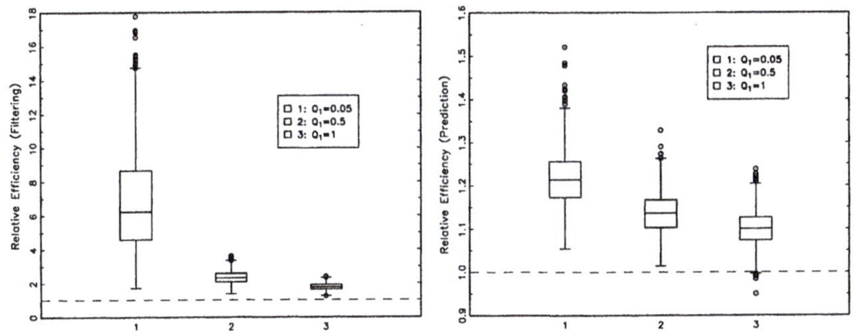

Fig. 7.15. Relative efficiency of the Kalman filter vs. the AMF(1) for different values of Q_1, DGP with bimodal innovation density, filtering and prediction

Table 7.5. Relative efficiency of the AMF(1) vs. the Kalman filter for different values of Q_1, DGP with bimodal innovation density

	Filtering			Prediction			
#	Mean	Std Dev	frac	Mean	Std Dev	frac	Variant
1	6.839	2.862	1.000	1.218	0.065	1.000	$Q_1 = 0.05$
2	2.360	0.370	1.000	1.136	0.046	1.000	$Q_1 = 0.5$
3	1.783	0.191	1.000	1.102	0.039	0.995	$Q_1 = 1.0$

While we omitted the results of the AMF(3) in section 7.1 because they were very similar to those of the AMF(1), we observe more substantial differences between the two for the scenario considered here. As apparent from table 7.6 and figure 7.16, the gain in efficiency is only small for the prediction problem. For the filtering problem, however, the AMF(3) shows some improvement over the AMF(1).

7.2.4 Estimation of Hyperparameters

Analogously to the simulation from section 7.1.4, hyperparameters are estimated for the model with bimodal distribution of the state innovation. Again, 1000 time series of length 350 are generated based on the data generating

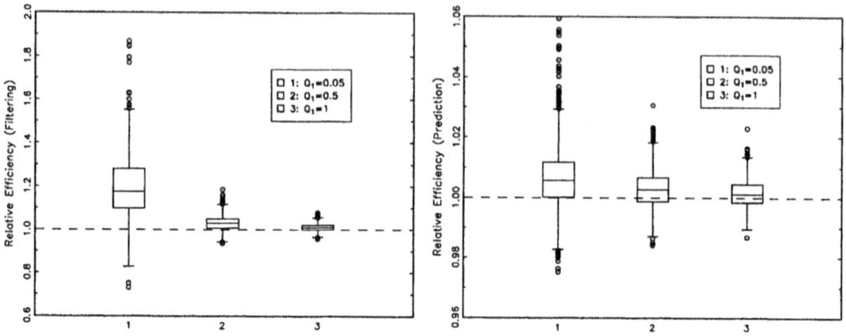

Fig. 7.16. Relative efficiency of the AMF(1) vs. the AMF(3) for different values of Q_1, DGP with bimodal innovation density, filtering and prediction

Table 7.6. Relative efficiency of the AMF(3) vs. the AMF(1) for different values of Q_1, DGP with bimodal innovation density

	Filtering			Prediction			
#	Mean	Std Dev	frac	Mean	Std Dev	frac	Variant
1	1.193	0.142	0.945	1.007	0.011	0.756	$Q_1 = 0.05$
2	1.031	0.035	0.808	1.003	0.006	0.675	$Q_1 = 0.5$
3	1.011	0.018	0.742	1.001	0.005	0.615	$Q_1 = 1.0$

process (7.14) - (7.21), setting the parameter values according to (7.22). We treat the model parameters μ_1, μ_2, Q_1, ω_1, T and h as unknown. For each generated time series, they are estimated using the approximate likelihoods based on the AMF(1) and the AMF(3). For the estimation we impose the restriction that

$$\omega_1\mu_1 + (1-\omega_1)\mu_2 = 0.$$

This leaves us with μ_1, Q_1, ω_1, T and h to estimated. These model parameters are reparameterized as:

$$T = 1 - \left(\frac{\psi_1}{10}\right)^2, \quad \omega_1 = \left(\frac{\psi_2}{10}\right)^2, \quad Q_1 = \left(\frac{\psi_3}{5}\right)^2, \quad h = (2\psi_4)^2, \quad \mu_1 = \psi_5^2. \tag{7.23}$$

The vector ψ^* that corresponds to the true parameter values from (7.22) is

$$\psi^* = (3.162,\ 3.162,\ 5.0,\ 2.236,\ 3.0)',$$

starting values for the optimization are set to

$$\psi^0 = (3.0,\ 3.0,\ 4.0,\ 3.0,\ 3.0)'.$$

The reparameterization imposes positivity restrictions on the variances Q_1 and h as well as on the weight ω_1. Furthermore, the mean reversion parameter T is bounded by 1 from above.

We compute the overall variance $\hat{Q}$ of the state innovation implied by the estimated mixture components $\hat{\omega}_1$, $\hat{\mu}_1$ and $\hat{Q}_1$ as

$$\hat{Q} = \hat{\omega}_1(\hat{Q}_1 + \hat{\mu}_1^2) + (1 - \hat{\omega}_1)(\hat{Q}_1 + \hat{\mu}_2^2)$$

Again, the quasi-likelihood based on the Kalman filter does not allow to estimate the components of the innovation mixture separately, so we estimate the innovation variance Q as a whole only. Model parameters are reparameterized as

$$T = 1 - \left(\frac{\psi_1}{10}\right)^2, \quad Q = \psi_2^2, \quad h = (2\psi_3)^2 \tag{7.24}$$

implying

$$\psi^* = (3.162,\ 3.162,\ 2.236)$$

as the value of ψ corresponding to (7.22). As starting values for the optimization procedure we choose

$$\psi^0 = (3.0,\ 3.0,\ 3.0)'.$$

Simulation results are provided by figures 7.17 and 7.18 and table 7.7. We first consider the model parameters T, Q and h that have been estimated by all three algorithms. The bias for the three parameter estimates is smaller than 0.5% for all algorithms considered. The estimates from the AMF show a smaller standard deviation than the Kalman filter. The standard deviations for the AMF(1) and the AMF(3) estimates are nearly equal. However, here as well as for the estimated ω_1, Q_1 and μ_1, the standard deviation of the AMF(3) never exceeds that of the AMF(1).

Concerning filtering with estimated parameters, the mean efficiency gain of the AMF(1) over the Kalman filter is approximately 62.2% which is less than the average gain of 78.3% observed for the corresponding case with known parameters.[15] In all of the 1000 runs, the MSE corresponding to the AMF(1) has been smaller than that corresponding to the Kalman filter. The mean efficiency gain from the AMF(3) over the AMF(1) is very small (0.6%), the MSE of the AMF(3) has been smaller than that of the AMF(1) for 65.7% of the runs.

The components ω_1 and μ_1 of the state innovation's mixture distribution are estimated with a bias below 0.5%, the bias of the variance component Q_1 however is quite high with 7.7% for the AMF(1) and 4.0% for the AMF(3). The boxplots in figure 7.18 suggest quite symmetric distributions for $\hat{\omega}_1$ and $\hat{\mu}_1$ while the distribution of $\hat{Q}_1$ appears to be skewed to the right.

Figure 7.19 shows as a solid line the true density of the state innovation η_t for the parameterization (7.22). The short-dashed line and the long-dashed line show average estimated innovation density obtained from using the AMF(1) and the AMF(3) respectively. These densities are defined in an

[15] See table 7.5 above.

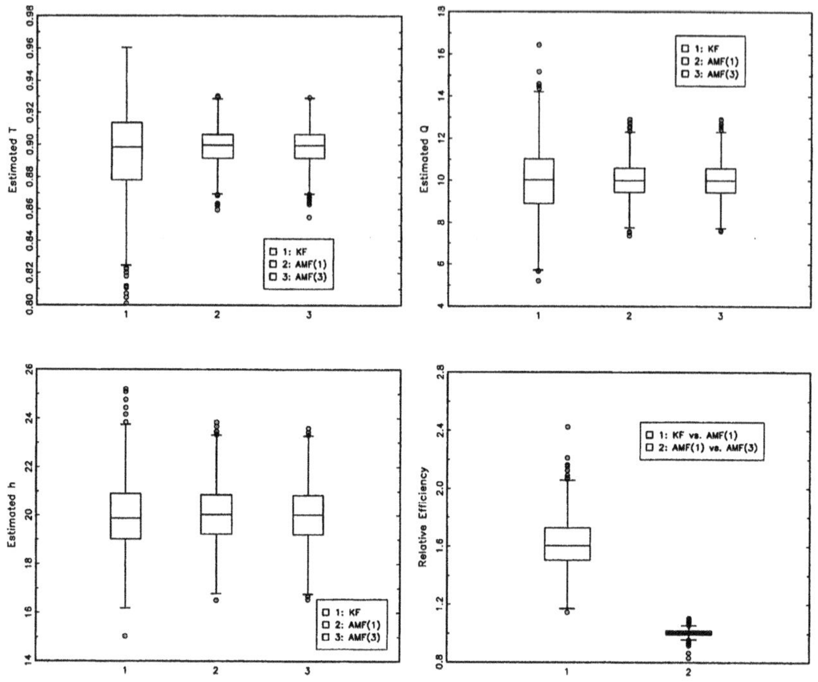

Fig. 7.17. Distribution of parameter estimates (1) and relative efficiency, DGP with bimodal innovation density

Table 7.7. ML estimates of parameters, DGP with bimodal innovation density

		Mean of estimate			Standard deviation		
Parameter	true	KF	AMF(1)	AMF(3)	KF	AMF(1)	AMF(3)
T	0.9	0.895	0.899	0.899	0.026	0.011	0.011
Q	10.0	10.031	10.044	10.031	1.595	0.878	0.872
h	20.0	19.977	20.055	20.046	1.422	1.247	1.228
ω_1	0.10	-	0.100	0.100	-	0.007	0.006
Q_1	1.0	-	0.923	0.960	-	0.297	0.275
μ_1	9.0	-	9.043	9.023	-	0.394	0.386

analogous manner to (7.12) above. The bimodal shape of the true innovation density is captured well by both variants of the AMF. By its very nature the average pdf implied by the Kalman filter is not capable of generating the bimodal shape.

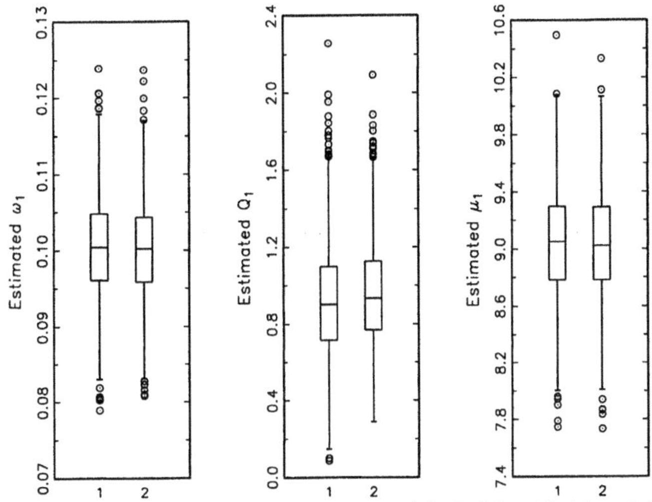

Fig. 7.18. Distribution of parameter estimates (2), DGP with bimodal innovation density, (left boxplot: Kalman filter, right boxplot: AMF(1))

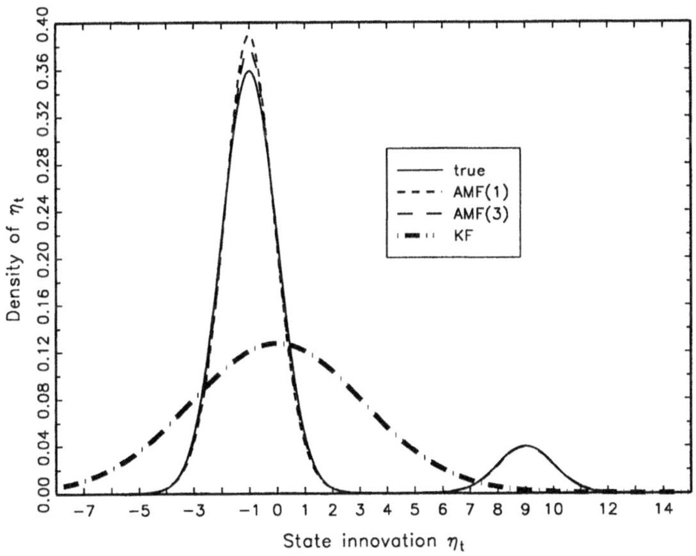

Fig. 7.19. True and estimated 'average' density of the state innovation, DGP with bimodal innovation density

7.3 Sampling from a Student t Distribution

7.3.1 Data Generating Process

The last simulation to be presented in this chapter is different in nature from the simulations considered up to here, insofar as the state innovation is not a Gaussian mixture anymore. It is rather specified as a Student t distribution with $v = 3$ of degrees of freedom. The structure of the rest of the data generating process is the same as in the previous sections.

For the transition equation we have

$$\alpha_t = T\alpha_{t-1} + \eta_t, \tag{7.25}$$

with

$$\eta_t \sim \text{ the Student } t \text{ distribution with } v = 3 \text{ degrees of freedom} \tag{7.26}$$

For $v = 3$ the mean and the variance of the t distribution exist and we have

$$E(\eta_t) = 0, \quad Var(\eta_t) = \frac{3}{3-2} = 3 =: Q.$$

The fourth moment $E(\eta_t^4)$, however, does not exist in this case.

The measurement equation is again given by

$$\begin{pmatrix} y_{1t} \\ y_{2t} \end{pmatrix} = Z\alpha_t + \epsilon_t \tag{7.27}$$

with

$$\epsilon_t \sim N(0, H), \tag{7.28}$$

where

$$H = h \cdot I_2 =: r \cdot Q \cdot I_2. \tag{7.29}$$

ϵ_t and η_t are each serially independent and independent from each other.

The state process is initialized by drawing α_0 from the student t distribution with 3 degrees of freedom. The following values are assigned to the model parameters:

$$T = 0.5, \quad Z = \begin{pmatrix} 2 \\ -1 \end{pmatrix}, \quad r = \tfrac{5}{3}. \tag{7.30}$$

A typical realization from this state space model is shown in figure 7.20. By the very nature of the t distribution with a small number of degrees of freedom, realizations of state innovations can be very extreme.

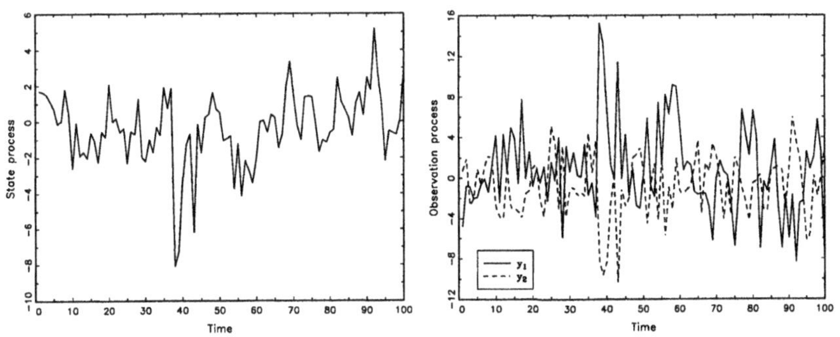

Fig. 7.20. Typical realization of the state and observation process, DGP with innovation from t distribution, $T = 100$ observations

7.3.2 Estimation of Hyperparameters

Since this time the state innovation is not distributed as a Gaussian mixture, we cannot carry out those exercises from the two sections above in which we conducted filtering and prediction with known parameters.

For the simulation here, we assume a situation where the statistician correctly assumes the functional form of the transition and the measurement equation but does not know the true nature of the distribution of the state innovation. The model parameters to be estimated are T and h. Parameter estimation is conducted first under the assumption of a Gaussian state innovation and second under the assumption of a mixture of the type (7.2). If interest is only in T and H, the parameters of the (incorrectly specified) distribution of the state innovation can be considered as nuisance parameters. Assuming normality one has to estimate the variance Q only. Under the assumption of a Gaussian mixture, one has to estimate ω_1, Q_1 and Q_2.

Again, the simulation is based on 1000 runs. For each run, a time series of 370 observations is generated whose first 20 observations are discarded. Under the assumption of a mixture distribution, parameters are estimated using the AMF(1) algorithm.[16] The model parameters are reparameterized as

$$T = \frac{\psi_1}{10}, \quad \omega_1 = \left(\frac{\psi_2}{10}\right)^2, \quad Q_1 = \psi_3^2, \quad Q_2 = \left(\frac{\psi_4}{5}\right)^2, \quad h = \left(\frac{\psi_5}{5}\right)^2 \tag{7.31}$$

The true model parameters imply $\psi_1 = 5.0$ and $\psi_5 = 11.180$. There are no true values corresponding to ψ_2, ψ_3 and ψ_4, since the true model distribution is not a Gaussian mixture. For each run we choose as starting values

$$\psi^0 = (5.0,\ 4.0,\ 7.0,\ 6.0,\ 11.0)'.$$

[16] Results from using the AMF(3) are not shown, since they turned out to be highly similar to those from the AMF(1).

For the estimation based on the Kalman filter the following reparameterization is chosen:

$$T = \frac{\psi_1}{10}, \quad Q = \psi_2^2, \quad h = \left(\frac{\psi_3}{5}\right)^2. \tag{7.32}$$

Again we have $\psi_1 = 5.0$ and $\psi_3 = 11.180$ corresponding to the true model parameters. Furthermore, the variance Q of the normal distribution corresponds to the variance of the student t distribution if $\psi_2 = 1.732$. The estimation process is started with

$$\psi^0 = (5.0,\ 2.0,\ 11.0)'.$$

The results in table 7.8 show that the variance of the measurement error is estimated with a small bias (<0.5%) only, whereas $\hat{T}$ and $\hat{Q}$ exhibit a bias around 1.5%.[17] The standard deviation of $\hat{h}$ and $\hat{T}$ can be judged as being satisfactory, whereas the standard deviations of $\hat{Q}$ is quite high for the Kalman filter as well as for the AMF(1). The corresponding boxplot, the second panel in figure 7.21, shows the occurence of one very extreme observation of $\hat{Q}$ for both algorithms.[18] Deleting these points from the two sets of estimated $\hat{Q}$, the standard deviation reduces to 1.214 for the Kalman filter and to 1.206 for the AMF(1).

Table 7.8. ML estimates of parameters, DGP with Student-t innovation density

		Mean of estimate		Standard deviation	
Parameter	true	KF	AMF(1)	KF	AMF(1)
T	0.500	0.492	0.494	0.060	0.051
Q	(3.0)	2.955	2.950	1.641	1.524
h	5.0	5.018	5.019	0.361	0.357
ω_1	-	-	0.120	-	0.113
Q_1	-	-	41.549	-	124.646
Q_2	-	-	1.294	-	0.471

Thus, it can be concluded that T and h can be estimated with acceptable precision, although wrong models for the state innovation have been assumed. For the AMF(1) and the Kalman filter, the quality of the estimators is similar, with the AMF(1) having a small advantage with respect to the standard deviation.

Filtering based on estimated parameters has also been conducted. Using the AMF(1) instead of the Kalman filter yields an average efficiency gain of

[17] For Q the true value has been set in parentheses. This is because Q is not a parameter of the true underlying model. Rather, Q is the variance of the falsely assumed distribution of the state innovation, and Q=3 implies that η_t has the same variance in the true model as in the auxiliary models.

[18] For both algorithms, this extreme observation showed up for the same run.

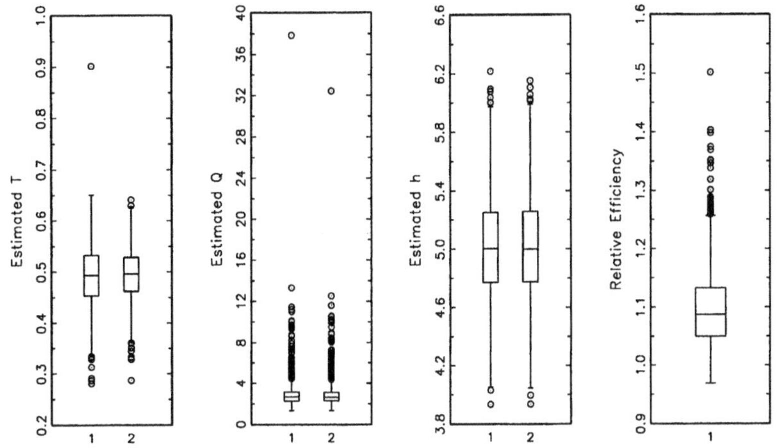

Fig. 7.21. Distribution of parameter estimates, DGP with innovation from t distribution (left boxplot: Kalman filter, right boxplot: AMF(1)) and relative efficiency.

10.0 %. Over the 1000 runs, the MSE obtained from the AMF(1) has been smaller than the one from the Kalman filter in 97.1% of all runs.

As indicated by the boxplots in figure 7.22, the possibility of extreme realizations is even more prominent for the individual components of the estimated mixture distribution.[19] While the second variance component $\hat{Q}_2$ takes on 'moderate' values only, $\hat{Q}_1$ assumes five values that are bigger than 600. It is interesting to explore under what circumstances these very extreme estimates occur. Moreover, it may be questionable if these values really correspond to a (global) maximum of the criterion function or if they rather occur due to some numerical problems.

Looking at the data that give rise to such extreme estimates more closely, one observes that they contain some very extreme observations which, however, are perfectly compatible with the t distribution used in our model. Figure 7.23 shows such an instance.

The estimates corresponding to this time series are given as

$$\hat{T} = 0.467,\ \hat{Q} = 6.374,\ \hat{h} = 5.407$$

for the Kalman filter and as

$$\hat{T} = 0.509,\ \hat{\omega}_1 = 0.005,\ \hat{Q}_1 = 1413.548,\ \hat{Q}_2 = 2.224,\ \hat{h} = 5.569,\ \hat{Q} = 8.725$$

for the AMF(1). While the estimate of the variance component Q_1 is very high, the corresponding estimate of the weight ω_1 is very low, leading to an

[19] The two boxplots in the middle both describe the distribution of $\hat{Q}_1$. The right of the two boxplots ignores realizations over 200, i.e. it 'zooms in' the left side of the distribution.

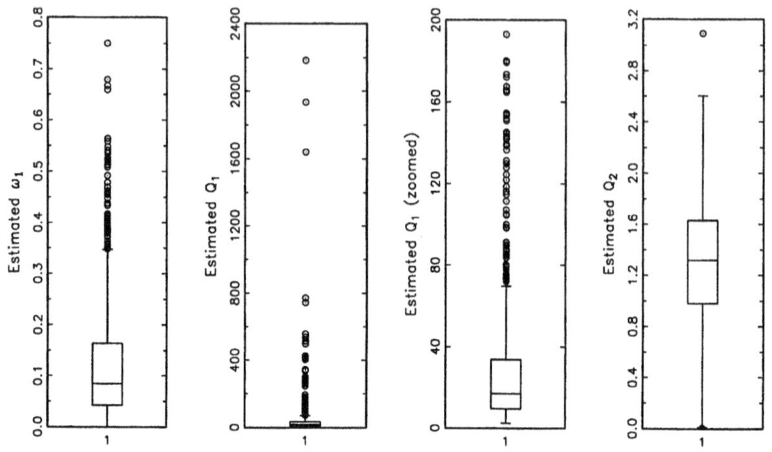

Fig. 7.22. Distribution of parameter estimates (2) obtained from AMF(1), DGP with innovation from t distribution

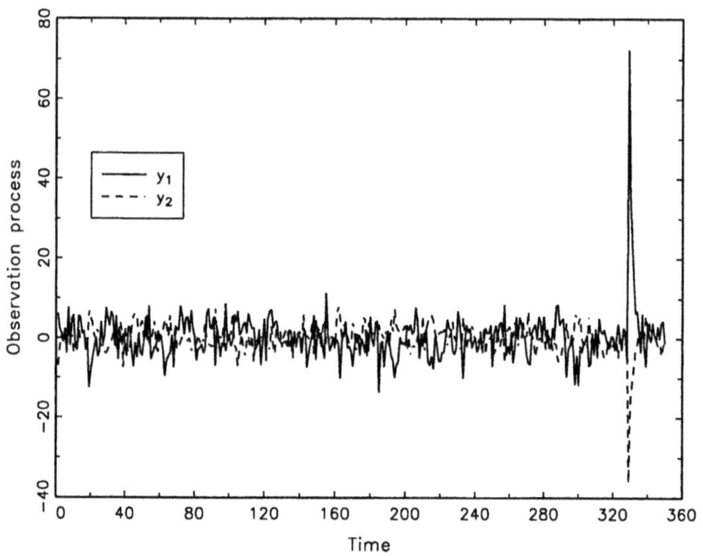

Fig. 7.23. A realization of the observation process containing a very extreme observation, DGP with innovation from t distribution

estimated overall variance $\hat{Q}$ of moderate size. This result is quite intuitive since it adequately captures the observed phenomenon of the data from the viewpoint of a normal mixture: highly extreme observations are allowed to occur, but only rarely.

As for potential problems with numerical optimization, we can ensure that the approximate likelihood based on the AMF(1) has in fact a local maximum at $\hat{\psi}$.[20] Moreover, the optimization algorithm has been run for different starting values all of which yield approximately the same estimator. This suggests that the estimate in fact corresponds to a global maximum.

Finally, the true density from the Student t distribution with $v = 3$ degrees of freedom is compared with the averaged estimated densities. Figure 7.24 shows that the average implied mixture distribution gives a much better approximation to the true density of the state innovation than the average implied normal distribution.

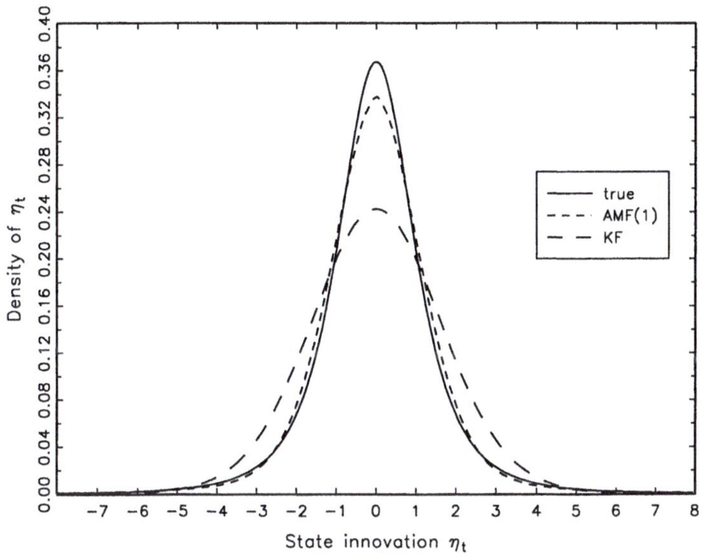

Fig. 7.24. True and estimated 'average' density of the state innovation, DGP with innovation from t distribution

7.4 Summary and Discussion of Simulation Results

In general it is not possible to compare the properties of the AMF, the exact filter, and the optimal linear filter analytically. Therefore, simulations have been conducted in this section in order to win an impression of the relative

[20] This has been checked as follows: For each component k of ψ, the likelihood has been drawn as a function of ψ_k only while keeping all other components of $\hat{\psi}$ fixed. The resulting functions all had one maximum corresponding to $\hat{\psi}_k$.

performance of the different algorithms for filtering, prediction and parameter estimation. Although simulations can never *prove* any properties of the estimation methods considered, they can nevertheless highlight certain key characteristics. The simulations also show that certain methods can in fact be practically applied to problems of interest: for instance, it has not been clear a priori, that numerical maximization of the approximate likelihood based on the AMF can be conducted without trouble.

The simulation results show that the AMF(1) and the AMF(3) give a good approximation to the exact filter for both filtering and prediction. This result holds for computing the conditional mean and variance of the state as well as for computing the whole conditional density. In most cases the results from the AMF(1) and the AMF(3) hardly differ from each other. However, the AMF(3) shows a substantial gain in efficiency for the example with the bimodal mixture. For all situations considered, the AMF exhibits a smaller average MSE[21] than the Kalman filter.

The dependence of the efficiency gain on different parameterizations has also been examined. In the case of the unimodal mixture distribution, for instance, relative efficiency decreases with the mean reversion parameter T for filtering and increases in T for prediction. For both types of estimation problems, the efficiency gain rises with Q_1, the variance ratio of the two mixture components.

It has been shown that unknown hyperparameters can be estimated using the approximate likelihood based on the AMF. The mixture density implied by the parameter estimates captures the shape of the true model density quite well. Using the quasi-likelihood function based on the Kalman filter, it is not possible, to estimate the single components of the mixture density from the data generating process.

Using the AMF as opposed to the Kalman filter also has some advantage in the situation in which the state innovation is distributed as a Student t and not as a Gaussian mixture. The estimators of the mean reversion parameter T and the measurement error matrix H show approximately the same precision for the Kalman filter and for the AMF. Concerning filtering based on estimated parameters, however, for 97.1% of all runs, the AMF(1) exhibits a smaller MSE than the Kalman filter. The average estimated innovation density corresponding to the mixture model bears a closer resemblance to the true Student t distribution than the average estimated innovation density implied by the Kalman filter.

For the different types of estimation problems, the simulations in this section can only convey a first impression concerning the relative performance of the AMF. Conducting similar simulations for other versions of the state space model would probably yield additional useful insights. These may confirm the results obtained here but they may also give rise to converse conclusions.

[21] Averaged over the 1000 runs of a simulation.

Within our simulations, the AMF(k) results have been often compared to the results of the optimal linear filter, i.e. the Kalman filter. For future research it may be instructive to compare the AMF(k) to other approximate filters from the literature.

New simulations may use mixture densities with more than two components, nonstationary state processes, time series of different lengths, vector valued state processes, system matrices that depend on time or other covariables, etc. Moreover, the results of any simulation can be sharpened by using more than the 1000 replications applied here.

8

Estimation of Term Structure Models in a State Space Framework

In this chapter we show how to use the statistical framework introduced in chapters 5 and 6 for estimating the term structure models presented in chapters 3 and 4. We describe how the theoretical models can be cast into state space form, i.e. how the transition and measurement equation are specified in accordance with the theoretical model. After that, suitable estimation techniques are presented. It is described how diagnostic checking can be conducted and what interpretation can be given to estimation results. An empirical study with discrete-time multifactor models will be conducted in chapter 9 below.

The term structure models that we have discussed above have been characterized by a factor process $\{X_t\}$ that drives the whole term structure. That is, at each time t, for an arbitrary collection of maturities $\{n_1, \ldots, n_k\}$, the corresponding yields $(y_t^{n_1}, \ldots, y_t^{n_k})$ depend on the factor vector at time t. The factor vector has been treated as latent. We have discussed discrete-time models, in which the factor evolution is a vector autoregressive process, and continuous-time models, in which the factor evolution is specified by a stochastic differential equation.

Consider now the problem of estimating a particular term structure model. We assume that a sample $\{y_1, \ldots, y_T\}$ of observations[1] of zero coupon yields is available. For each t, the vector $y_t = (y_t^{n_1}, \ldots, y_t^{n_k})'$ consists of observations of k different yields. For estimating a given theoretical term structure model and for assessing how it fits the empirical data, one may distinguish between three types of approaches.

First, there are studies that analyze whether the dynamic specification of a theoretical model is adequate. The most important examples for this approach are the numerous studies that estimate time series models for the short rate, the latter being interpreted as the single factor in one-factor models.[2] This

[1] In the present context we do not distinguish between truly observed data and those that have been constructed, compare the discussion in section 2.2.

[2] See, e.g. [29] as a frequently cited example. - Econometric analysis of the short rate evolution has of course its own independent value. That is, the analysis can

is an instructive exercise, since it analyzes whether the chosen specification of the short rate evolution is suitable. However, by its very nature it can tell nothing about whether the term structure model based on this short rate process is successful in capturing the cross-section properties of yields.

A second type of approach uses cross-section information only. Let the theoretical solution for bond prices be a function of g parameters and the d-dimensional unobservable factor vector at time t. Given the observation of bond prices for $k > g + d$ maturities at time t, estimates of parameters and the unobservable state can be found by minimizing a distance criterion between theoretical and observed prices. This approach is employed by [25] for estimating the model by [34]. A disadvantage of the pure cross-section approach is that the parameter estimates may change over time. Even if they are constant, the estimated factor process - implied by estimating the model at each point in time separately - needs not to be in accordance with the model of interest.[3]

Both approaches - using solely time series or solely cross section information - do not exploit the full information which is provided by the shape of the term structure at a given point in time *and* by its dynamic evolution. In light of the drawbacks of these approaches, a third type of approach estimates term structure models using the whole panel of data.

A problem arising with this approach is the unobservability of factors. One solution to that problem consists of working with data for which $k = d$, i.e. the number of yields at each time t is equal to the number of factors. Then the term structure equation (i.e. yields as a function of factors) can be inverted, and the dynamic specification of the factor process implies a multivariate time series model for the evolution of observable yields. The problem with this approach is, that it restricts the number of yields to be used in the cross section. Moreover, it is quite arbitrary which yields (i.e. which maturities) are selected for estimation, and parameter estimates may possibly vary depending on which collection of maturities has been chosen.

In the literature, the state space approach has been adopted for the estimation of term structure models, since it removes some of the problems associated with the other approaches outlined above.[4] For estimating a term structure model in state space form, one has to transform the factor evolution to the form of a state space model's transition equation. For affine discrete-time term structure models this is a straightforward exercise. In most cases the factor process of the theoretical model already has that required structure. For continuous-time models, the factor process has to be brought into discrete time. The measurement equation basically arises by choosing observed prices or interest rates as left-hand-side variables, whereas the right-hand-side is the

---

also be conducted without having a fully specified arbitrage-free dynamic term structure model in mind, of which the analyzed short rate process is an ingredient.

[3] See [22].

[4] An overview of the literature will be given below.

sum of the theoretical solution implied by the term structure model and a measurement error.

Once a theoretical term structure model is cast into state space form, the way is open for conducting statistical inference using the methodology (filtering, estimation, testing) available for state space model analysis. For instance, when the implied state space model is linear and Gaussian, the optimal estimation technique is maximum likelihood based on the Kalman Filter.

The advantages of the state space approach for the estimation of term structure models can be summarized as follows: one does not have to rely on a proxy for the state process; the unobservable state process can be estimated; the market price of risk parameters can be identified; the specification of the factor process according to the model of interest can be accounted for; the bond price formulae implied by the no-arbitrage condition are integrated; and one can explicitly account for measurement errors driving a wedge between theoretical and observed bond prices or yields.

8.1 Setting up the State Space Model

In the following we will describe how the theoretical models can be cast into state space form, i.e. how the transition and measurement equation are specified in accordance with the theoretical model.

8.1.1 Discrete-Time Models from the AMGM Class

For the multifactor models in discrete time, setting up a corresponding state space model is in many cases straightforward. Let us assume that we are faced with a d-factor model[5] from the AMGM class in canonical form, whose factor process is given by[6]

$$X_t = \mathcal{K} X_{t-1} + u_t \tag{8.1}$$

with

$$u_t \sim \sum_{b=1}^{B} \omega_b N(0, V_b). \tag{8.2}$$

Absence of arbitrage implies that the yield y_t^n of a bond with maturity n is given by

$$y_t^n = \frac{A_n}{n} + \frac{1}{n} B_n' X_t, \tag{8.3}$$

[5] Note that the character d is used for both, the dimension of the factor vector in the term structure model and for the intercept vector in the measurement equation of the state space model. However, it will be clear from the context which is referred to.

[6] These are the same equations as in chapter 3 below. They are repeated here for convenience. Moreover, we set $\mu_b = 0$.

where A_n and B_n depend on the parameters of the factor process and on additional market price of risk parameters collected in a vector λ.

Suppose we are given observations of a multiple time series $\{y_1, \ldots, y_T\}$ where each y_t contains the yields for a set $(n_1, \ldots, n_k)'$ of maturities, i.e. $y_t = (y_t^{n_1}, \ldots, y_t^{n_k})'$. We want to put the theoretical model into state space form, so that the latter can be used as a vehicle for conducting statistical inference.

A natural choice for the state of the system is the vector of factors itself, i.e. we set $\alpha_t = X_t$. This implies that the transition equation for the state vector is linear,

$$\alpha_t = c + T\alpha_{t-1} + \eta_t, \quad \eta_t \sim \sum_{b=1}^{B} \omega_b N(0, Q_b).$$

and the form of the system matrices can be read off one to one from the law (8.1) of the factor evolution. We have $c = 0$, $T = \mathcal{K}$, $\eta_t = u_t$, and $Q_b = V_b$.[7]

As for the measurement equation, the measurement vector is given by the k-dimensional vector of yields. The theoretical model implies that

$$\begin{pmatrix} y_t^{n_1} \\ \vdots \\ y_t^{n_k} \end{pmatrix} = \begin{pmatrix} \frac{1}{n_1} A_{n_1} \\ \vdots \\ \frac{1}{n_k} A_{n_k} \end{pmatrix} + \begin{pmatrix} \frac{1}{n_1} B'_{n_1} \\ \vdots \\ \frac{1}{n_k} B'_{n_k} \end{pmatrix} \alpha_t \tag{8.4}$$

Adding a vector of measurement errors leads to a linear measurement equation

$$y_t = d + M\alpha_t + \epsilon_t, \quad \epsilon_t \sim N(0, H) \tag{8.5}$$

with obvious definitions of the vector d and the matrix M.

Thus, we have the intermediate result that the evolution of a vector of yields can be represented by a linear state space model if the underlying theoretical model is from the AMGM class. Moreover, the transition equation coincides with the VAR(1) process of the factor evolution. Obviously, for the case $B = 1$, the resulting state space model is a linear Gaussian model. If $B > 1$ it is a state space model for which the state innovation is distributed as a Gaussian mixture, and which has been discussed in chapter 6 above.

The implied state space model retains its simple structure if term spreads are included in the measurement vector y_t. Denote by $s_t^{m,n} := y_t^n - y_t^m$ the spread between the n-month yield and the m-month yield. Then the theoretical model implies that

$$s_t^{m,n} = \left(\frac{A_n}{n} - \frac{A_m}{m}\right) + \left(\frac{B_n}{n} - \frac{B_m}{m}\right)' X_t.$$

[7] Note that for the weight parameter in the mixture distribution, we use the same symbol ω_b for state space models as well as for term structure models.

So, if $s_t^{m,n}$ is the ith entry in measurement vector, the ith element in the d vector and the ith row of the M matrix in the corresponding state space model are given as

$$\left(\frac{A_n}{n} - \frac{A_m}{m}\right) \text{ and } \left(\frac{B_n}{n} - \frac{B_m}{m}\right)'$$

respectively. The state process does still coincide with the factor vector.

One could also include yields in first differences into the observation vector. This, however, implies that the state process has to be adjusted. For $\Delta y_t^n = y_t^n - y_{t-1}^n$, the model implies

$$\Delta y_t^n = \frac{1}{n} B_n' \Delta X_t.$$

Having Δy_t^n as the ith entry in the measurement vector, it cannot be expressed as a function of the current state α_t. This is because it depends also on X_{t-1} which is not part of α_t. However, we can come up with a new state vector that incorporates X_{t-1}. Accordingly, we must specify a new transition equation that has still a Markovian structure. So if we have yield changes in the observation vector, we define the new state vector as $\alpha_t = (X_t', X_{t-1}')'$. Its transition equation can then be written as[8]

$$\alpha_t = \begin{pmatrix} \mathcal{K} & 0 \\ I & 0 \end{pmatrix} \alpha_{t-1} + \begin{pmatrix} u_t \\ 0 \end{pmatrix}.$$

For the measurement equation, the M matrix has to be enhanced by d columns. The ith element in the vector d and the ith row of the matrix M are given as

$$0 \text{ and } \left(\frac{B_n'}{n}, -\frac{B_n'}{n}\right)$$

respectively.

For the discrete-time AMGM models, we have seen that including yields in levels, spreads, or yield changes into the measurement vector leads to a state space model which is linear. This stems from the fact that yields are affine in factors. If we want to set up the state space model for observable financial quantities which are not affine functions of yields, the linearity of the measurement equation breaks down.

8.1.2 Continuous-Time Models

Having discussed discrete-time models from the AMGM class, we now turn to the question how continuous-time models can be cast into state space form.

[8] This implies that the transition matrix T does not have full rank and that the innovation vector has a degenerate distribution. For estimation purposes, however, that does not lead to real problems.

We first consider models from the exponential-affine class, introduced in section 4.2. For these models, the formulation of the measurement equation is straightforward, if the vector of measurements contains yields. Equation (4.22) shows that yields are affine functions of the factors. Note that this is the same structure as for the discrete-time models. Thus, if the factors are included in (or identical to) the state vector, the measurement equation for continuous-time models from the exponential-affine class has the same structure as (8.4). Of course, however, the model parameters enter the functions A_n and B_n in a different way.

For continuous-time models, the difficult part is the translation from the stochastic differential equation (SDE) of the factor process, to a discrete-time transition equation. Consider a time interval $R = [0, T^*]$ for which a particular model from the exponential-affine class is supposed to hold. Within this interval there are T equidistant points of measurement. Let the set of points where the measurement is taken be given by $\{t_1, t_2, \ldots, t_T\}$.

In order to specify the discrete-time transition equation, the continuous-time SDE has to be discretized. That is, given the specification of the factor process $\{X_t, t \in R\}$ as an SDE and some $h > 0$ the task is now to derive a relationship between the two random variables X_t and X_{t+h}. We will describe an approach that is based on solving the SDE.[9]

One starts by solving the SDE such that X_{t+h} is explicitly expressed in terms of X_t. Then a transition density $p(X_{t+h}|X_t)$ is derived. If possible, this is in turn used to construct a transition equation of the form

$$\alpha_t = c + T\alpha_{t-1} + \eta_t, \quad \eta_t \sim D(0, Q), \tag{8.6}$$

where D denotes an arbitrary distribution. For models of the exponential-affine class it can be shown that the transition density is Gaussian when the volatility is constant; i.e. the matrix consisting of the β_i in (4.24) is the zero matrix:

$$\mathcal{B} = (\beta_1, \ldots, \beta_n) = 0.$$

In the case of an unrestricted $\mathcal{B}$, it is possible that either no closed form for $p(X_{t+h}|X_t)$ can be found or the transition equation is of a form that makes the estimation of the model very complicated. In the one-factor model by Cox, Ingersoll and Ross, [34], for example, the transition density can be shown to be a noncentral χ^2 density.[10]

In the Gaussian case - arising from $\mathcal{B} = 0$ - the transition density is fully described by the conditional mean function $m(X_t, h)$ and the conditional variance-covariance matrix $V(X_t, h)$. We anticipate a result from below, where it is shown that this conditional variance-covariance matrix does not depend on the factors but only on the length h of the time interval between X_{t+h} and X_t. Accordingly, one can write

9 For other discretization approaches in this context, see, e.g., [12] and [16].

10 See [34]. The process for the short rate in that model is given by equation (4.34) above.

$$X_{t+h} = m(X_t, h) + \eta_{t+h}, \quad \eta_{t+h} \sim N(0, V(h)).$$

We will now show how to obtain the conditional mean $E(X_{t+h}|X_t)$ and the conditional variance-covariance matrix $Var(X_{t+h}|X_t)$ for the factor process of the type (4.29),

$$dX_t = \mathcal{K}(\theta - X_t) + \Sigma\sqrt{S_t}dW_t, \tag{8.7}$$

which is written here again for convenience.

The conditional mean and variance for the Gaussian case will then follow from setting $\mathcal{B} = 0$. We will assume that $\mathcal{K}$ is diagonal.[11] In order to justify this assumption for an exponential-affine model, one may again start with a model with arbitrary $\mathcal{K}$ and then apply affine transformations that lead to diagonality of $\mathcal{K}$.

In order to express X_{t+h} in terms of X_t one has to solve the SDE above.[12] We start by defining

$$Z_t = e^{\mathcal{K}t}X_t.$$

Using (8.7), we can obtain an integral representation of Z_t via Ito's lemma:

$$Z_t = \int_0^t e^{\mathcal{K}v}\mathcal{K}\theta dv + \int_0^t e^{\mathcal{K}v}\Sigma\sqrt{S_v}dW_v. \tag{8.8}$$

Now let $s > t$. From (8.8) it is easily observed that

$$Z_s = Z_t + \int_t^s e^{\mathcal{K}v}\mathcal{K}\theta dv + \int_t^s e^{\mathcal{K}v}\Sigma\sqrt{S_v}dW_v. \tag{8.9}$$

Multiplying through by $e^{-\mathcal{K}s}$, solving the deterministic integral and collecting terms yields:

$$X_s = \theta + e^{-\mathcal{K}(s-t)}(X_t - \theta) + \int_t^s e^{-\mathcal{K}(s-v)}\Sigma\sqrt{S_v}dW_v. \tag{8.10}$$

The conditional expectation of the stochastic integral in (8.10) is zero which follows from the martingale property of stochastic integrals.[13] Setting $h = s - t$ we have for the conditional mean:

$$E(X_{t+h}|X_t) = \theta + e^{-\mathcal{K}h}(X_t - \theta) \tag{8.11}$$

The conditional variance-covariance matrix is defined as

$$Var(X_s|X_t) \tag{8.12}$$
$$= E([X_s - E(X_s|X_t)][X_s - E(X_s|X_t)]'|X_t) \tag{8.13}$$
$$= E\left(\left[\int_t^s e^{-\mathcal{K}(s-v)}\Sigma\sqrt{S_v}dW_v\right]\left[\int_t^s e^{-\mathcal{K}(s-v)}\Sigma\sqrt{S_v}dW_v\right]'\middle| X_t\right) \tag{8.14}$$

---

[11] See [38].
[12] For the following see [78].
[13] See [19].

By a multivariate version of an Ito isometry result this becomes

$$Var(X_s|X_t) \tag{8.15}$$

$$= E\left(\int_t^s e^{-\mathcal{K}(s-v)} \Sigma S_v \Sigma' e^{-\mathcal{K}'(s-v)} dv | X_t\right) \tag{8.16}$$

$$= \int_t^s e^{-\mathcal{K}(s-v)} S_v E(\Sigma|X_t) \Sigma' e^{-\mathcal{K}'(s-v)} dv. \tag{8.17}$$

The typical (i, j)-element of this variance-covariance matrix is given by

$$[Var(X_{t+h}|X_t)]_{i,j} \tag{8.18}$$

$$= \frac{1 - e^{-(k_i+k_j)h}}{k_i + k_j}(a_{ij} + b'_{ij}\theta) + \sum_l \frac{(X_{tl} - \theta_l)b_{ijl}}{k_i + k_j - k_l}\left(e^{-k_l h} - e^{-(k_i+k_j)h}\right).$$

The scalars a_{ij} and the vectors b_{ij} depend on the elements of the model parameters Σ, α and $\mathcal{B}$. k_i is the ith diagonal element of the matrix $\mathcal{K}$.

From (8.11) and (8.18) we observe that both the conditional mean and the conditional variance-covariance matrix are affine functions of X_t. For the case $\mathcal{B} = 0$ we set additionally $\alpha = (1, \ldots 1)'$ without loss of generality, which implies $S_t = I$. For this case, the elements of the conditional variance-covariance matrix become

$$[Var(X_{t+h}|X_t)]_{ij} = \frac{1 - e^{-(k_i+k_j)h}}{k_i + k_j} a_{ij}. \tag{8.19}$$

The conditional mean remains unaffected by the specializing assumptions.

Let us turn back to the problem of finding a discrete-time transition equation for a state space model as a statistical representation of the continuous-time factor model. We have derived the conditional mean function $m(X_{t+h}, h)$ in (8.11) and the elements of the conditional variance-covariance matrix $V(h)$ in (8.19). Hence, we can write

$$X_{t+h} = \theta + e^{-\mathcal{K}h}(X_t - \theta) + \eta_{t+h}, \quad \eta_{t+h} \sim N(0, V(h))$$

where the i, j-element of $V(h)$ is given by (8.19). Now, take two consecutive points in our discrete time scale $\{t_1, \ldots, t_T\}$, say t_{k+1} and t_k, and choose h as the distance between the two, i.e. $h = t_{k+1} - t_k$. Identifying the state vector with the vector of factors, $\alpha_t = X_t$, we have the transition equation

$$\alpha_{t_{k+1}} = \theta - e^{-\mathcal{K}h}\theta + e^{-\mathcal{K}h}\alpha_{t_k} + \eta_{t_{k+1}}, \quad \eta_{t_{k+1}} \sim N(0, V(h)). \tag{8.20}$$

This is of the form (8.6) with $c = \theta - e^{-\mathcal{K}h}\theta$, $T = e^{-\mathcal{K}h}$, and $Q = V(h)$.

If the underlying theoretical model is characterized by level dependent volatility, i.e. $B \neq 0$, the distribution of X_{t+h} conditional on X_t is not Gaussian. However, in analogy to the Gaussian case, it is common in the literature to write an approximate transition equation which is based on the first two moments. Thus, setting $\alpha_t = X_t$ again, one writes

$$\begin{aligned} \alpha_{t_{k+1}} &= \theta - e^{-\mathcal{K}h}\theta + e^{-\mathcal{K}h}\alpha_{t_k} + \eta_{t_{k+1}} \\ \text{with } E(\eta_{t_{k+1}}|\alpha_t) &= 0, \;\; Var(\eta_{t_{k+1}}|\alpha_t) = V(\alpha_t, h) \end{aligned} \tag{8.21}$$

with i, j-element of $V(\alpha_t, h)$ given by (8.18).

8.1.3 General Form of the Measurement Equation

Before we come to an overview of the literature, we will have a closer look on the measurement equation of a term structure model in state space form. Let the general form of the measurement equation in a state space model

$$y_t = M_t(\alpha_t) + \epsilon_t \tag{8.22}$$

be decomposed into

$$g_Z(Z_t) = g_P(P_t^*) + \epsilon_t \tag{8.23}$$

$$P_t^* = g_\alpha(\alpha_t) \tag{8.24}$$

thus $y_t = g_z(Z_t)$ and $g_p(g_\alpha(\alpha_t)) = M(\alpha_t)$. Here Z_t is a vector of prices and interest rates that are actually observed in the market. The function $g_Z(\cdot)$ represents any transformation that the observed data is undergone before they are used for estimation. On the right hand side, P_t^* is the solution for zero bond prices as implied by the respective theoretical term structure model of interest. Via the function $g_\alpha(\cdot)$ this is written in terms of the state of the system which in turn depends on the factors of the model. Finally, $g_P(\cdot)$ writes the measurement variables in terms of zero bond prices.

Here are two representative examples. First, for many empirical studies in the literature the measurement vector y_t contains yields for different maturities, say $n_1, \ldots, n_k$. However, as outlined in section 2.2 above, these zero coupon yields are usually artificially constructed from observed market rates and prices such as interbank rates, futures, and swap rates. The construction methods can be parametric (approaches of Nelson-Siegel or Svensson) or non-parametric (spline smoothing). In our framework above, Z_t is the vector of all these market data that enter the estimation of artificial zero coupon rates at time t. The estimation procedure itself, e.g. the Nelson-Siegel method, is then captured by the mapping $g_Z(\cdot)$. If the model is of the exponential-affine class, zero bond prices are exponential functions of the factors. With $X_t = \alpha_t$, the function $g_\alpha(\cdot)$ transforms the state vector into the $k \times 1$ vector of theoretical zero bond prices. Finally, $g_P(\cdot)$ stands for the transformation (2.2) that maps prices into yields.

For a second example, consider the case of g_Z being the identity function. That is, market data are directly used as left-hand variables in the measurement equation. The vector y_t in [107], for instance, consists of observed LIBOR and swap rates of different maturities. Then the function $g_P(\cdot)$ in (8.23) represents the functional dependence between zero coupon prices and LIBOR (or swap) rates. The nature of the function $g_\alpha(\cdot)$ depends on the term structure model at hand. Due to the nonlinearity of $g_P(\cdot)$, the composed function $M = g_\alpha \circ g_P$ will be nonlinear in general.

The random vector ϵ_t in (8.23) represents the wedge between the observed y_t and its theoretical counterpart $M(\alpha_t)$ implied by the model. Within our general discussion of state space models, ϵ_t has been referred to as the measurement error. In the present context, the existence of this error term can be attributed, for instance to bid-ask spreads or errors in quoted prices. Moreover, if artificially constructed data (e.g. synthetic zero yields) are used, they are per se different from the true ones. Finally, ϵ_t captures all deviations of y_t from $M(\alpha_t)$ which are due to misspecification of the model.[14]

As for a specific distributional assumption, all studies in the literature assume

$$\epsilon_t \sim N(0, H).$$

Usually, the variance-covariance matrix H consists of constants. It is mostly specified as diagonal, frequently simply as $H = c \cdot I$, which is convenient for parameter estimation. Different elements for the diagonal may be chosen if one expects, e.g., that trading activity and bid-ask spreads vary across maturities.[15]

8.2 A Survey of the Literature

This section gives an overview of the literature that makes use of the state space model as a statistical framework for estimating arbitrage-free term structure models. Most of the literature deals with models that are formulated in continuous time.[16] There again, the majority of empirical studies is concerned with the estimation of models from the exponential-affine class.

The one-factor Vasiček model (4.33) is estimated, e.g., by [10], [9], [38], [40] and [97]. The one-factor CIR model (4.34) is investigated by, e.g., [12] and [40]. Estimation of multifactor versions of the CIR model (4.35) is the subject of articles by [55] and [52]. On a more theoretical level, the specific problems arising with estimating this class of models are discussed in [22]. The two-factor model by [15], (4.37), is estimated by [66], [78], [79], [55] and [97]. The extension of the CIR-model in which the evolution of the stochastic diffusion term $\sqrt{v_t}$ is itself governed by a square root process, (4.38)-(4.39), is dealt with by [97] and [103]. The three-factor model (4.40)-(4.42) by [30], that contains a process for the stochastic volatility term as well as a stochastic

[14] See [55].

[15] See [10].

[16] One recent example for estimating a discrete-time model using the state space methodology is [28]. They estimate a two-factor Gaussian model using German yield data.

central tendency for the short rate, is estimated by [8], a similar specification is examined by [97].

The two-factor model estimated by [70] formally belongs to the exponential -affine class as well. The peculiarity enters, however, in the way the market price of risk is specified. The model is written as

$$d\begin{pmatrix} X_{1t} \\ X_{2t} \end{pmatrix} = \left[\begin{pmatrix} \mu_1 \\ \mu_2 \end{pmatrix} + \begin{pmatrix} k_1 & 0 \\ k_2 & \theta_2 \end{pmatrix}\begin{pmatrix} X_{1t} \\ X_{2t} \end{pmatrix}\right] dt + \begin{pmatrix} \sigma_1 & 0 \\ 0 & \sigma_2 \end{pmatrix} d\begin{pmatrix} W_{1t} \\ W_{2t} \end{pmatrix},$$

where W_1 and W_2 are correlated $\mathbb{P}$-Brownian motions. The first factor has the interpretation of the short rate: $X_{1t} \equiv r_t$. As usual, W_1 and W_2 are transformed to Brownian motions under a risk neutral measure $\mathbb{Q}$ by the transformation (4.7). For W_2 the process λ_{2t} is chosen to be constant, $\lambda_{2t} \equiv \lambda_2$, thus $d\tilde{W}_{2t} = dW_{2t} + \lambda_2 dt$. The special feature of the model comes into play by choosing the process X_{2t} itself as the market price of risk process for r_t. That is, [70] sets $\lambda_{1t} \equiv X_{2t}$ yielding $d\tilde{W}_{1t} = dW_{1t} + X_{2t}dt$. From a technical point of view we have again an exponential-affine model with a factor process being affine under $\mathbb{P}$ as well as under $\mathbb{Q}$. The interesting point is that the market price of risk is itself one of the factors whose evolution through time is explicitly modeled. According to [70] this specification may be interpreted as the dynamic evolution of the preferences of a representative agent.

The studies by [39] and [16] employ a HJM approach for term structure modeling. As outlined above, the basic ingredient within the HJM approach is a specification of the volatility structure $\sigma(t, T)$ of forward rates.

The model of [39] contains two key variables, the instantaneous short rate r_t and ϕ_t that has the interpretation as an exponentially weighted moving average of the short rate's history. Under certain assumptions concerning the initial forward rate curve and the market price of risk specification, the joint dynamics of r_t and ϕ_t are given by

$$d\begin{pmatrix} r_t \\ \phi_t \end{pmatrix} = \left[\begin{pmatrix} a\bar{f} + \lambda\sigma_0^2 \\ \sigma_0^2 \end{pmatrix} + \begin{pmatrix} -(a - \lambda\sigma_1^2) & 1 \\ \sigma_1^2 & -2a \end{pmatrix}\begin{pmatrix} r_t \\ \phi_t \end{pmatrix}\right] dt + \begin{pmatrix} \sqrt{\sigma_0^2 + \sigma_1^2 r_t} dW_t \\ 0 \end{pmatrix} \tag{8.25}$$

This can be termed a degenerate case of a stochastic differential equation since there is only one source of randomness here. Nevertheless, from the system (8.25), (8.25), a linear discrete transition equation can be constructed as described above. The resulting state space model using yields in the measurement vector has linear measurement equation, since bond prices are exponential-affine in the states r_t and ϕ_t.

The HJM-model by [16] is based on the volatility structure

$$\sigma(t, T) = a_0 e^{-\lambda(T-t)} r_t^{\gamma}. \tag{8.26}$$

This leads to a three-dimensional diffusion of the form

$$dS_t = F(S_t; \psi)dt + V(S_t; \psi)d\tilde{W}_t. \tag{8.27}$$

The components of the state vector S_t have the following interpretation: S_{1t} is the logarithm of the price of a T-bond $P(t,T)$. S_{2t} is the short rate r_t. S_{3t} is a variable that captures the path history of the short rate. The drift and diffusion functions F and V are nonlinear in S_t. The model is transformed into state space form where the transition equation results from discretisizing (8.27). The one-dimensional measurement equation is constructed by extracting the log bond price from S_t and adding a noise term capturing measurement errors: $y_t = (1,0,0)S_t + \epsilon_t$.

Finally, [41] and [96] utilize the state space framework for estimating continuous-time multifactor models that incorporate default risk. [41] uses corporate bond yields, whereas [96] uses government bond yields from differently rated countries.

8.3 Estimation Techniques

For term structure models in state space form, let unknown parameters be collected in a vector ψ. These include the parameters showing up in the theoretical term structure model and the entries of the variance-covariance matrix of the measurement error in the measurement equation. In the following, we will outline which techniques are employed in the literature for estimating ψ.

If the state space model that is implied by a particular term structure model is linear and Gaussian, the optimal solution to the estimation problem is maximum likelihood estimation based on the Kalman filter. As a byproduct one also obtains optimally filtered or smoothed paths of the factor processes in the model. We have seen that a linear Gaussian state space model arises, if the underlying term structure model is a Gaussian multifactor model[17] in continuous or discrete time and the measurement vector contains (artificially constructed) zero bond yields. If one of these assumptions is dropped, the corresponding state space model will not exhibit that particularly convenient structure and the estimation techniques have to be adjusted.

One deviation from the linear Gaussian case has been extensively discussed in previous chapters: if in an affine term structure model, Gaussian factor innovations are replaced by innovations with a Gaussian mixture distribution, the resulting state space model can be estimated using the techniques presented in chapter 6. The following gives an overview of how other departures from linearity and normality are treated in the literature.

As mentioned above, the exponential-affine term structure model in continuous time with state dependent volatility, i.e. $\mathcal{B} \neq 0$, leads to a state space form in which the transition density is not normal any longer. An approximate affine transition equation is given by (8.21). Moreover, for this model

[17] This includes one-factor models as a special case.

two further peculiarities hold. First, the conditional variance-covariance matrix of η_t is an (affine) function of the previous period's state α_{t-1}. Second, by the specification of the diffusion matrix in (4.23) as the square root of an affine function of the factors, one can observe that negative states may cause a problem.

For the model with transition equation (8.21) and Gaussian measurement equation, the following approaches for filtering and parameter estimation have been proposed. All of them can be interpreted as modifications of the Kalman Filter algorithm and the corresponding maximum likelihood estimation, that has been described in chapter 5.

The prediction step is modified in [78], [38], [40] and [55] by replacing Q in (5.21) by $Q(a_{t-1})$. The latter has the form of the conditional variance-covariance matrix given in (8.21), but the unobservable α_{t-1} is replaced by the filtered estimate a_{t-1}. Using the estimated state for constructing the conditional variance-covariance matrix may lead to $Q(a_{t-1})$ becoming negative definite when an element of a_{t-1} is negative. In the literature two approaches are suggested in order to avoid this problem. One possibility is to replace the estimate a_t by zero, whenever it becomes negative. This is done by [38], [78] and [55]. The latter article alternatively proposes to replace a_t by a_{t-1} whenever a_t would become negative. Thus, a whole iteration step of the filter would be skipped in this case.

The conditional densities $p(y_t|\mathcal{Y}_t)$ (based on which the likelihood is constructed) are in general not Gaussian, when the transition density $p(\alpha_t|\alpha_{t-1})$ is not Gaussian. However, one can still construct quasi maximum likelihood (QML) estimates of ψ as mentioned in section 5.3.3 above.

In summary, the estimation procedure consists of the following components: Q in the Kalman Filter is replaced by $Q(a_{t-1})$; $a_t < 0$ is avoided; and an estimator ψ^* is constructed by maximizing the quasi-likelihood. An estimator which has been so constructed will be called a QML1 estimator hereafter.

As an alternative to the QML1 estimators, [78] offers the following method. Instead of using an estimator for the conditional variance-covariance matrix $Q(\alpha_t)$ he uses the unconditional variance-covariance matrix Q. Moreover, he ignores the nonnegativity restriction for a_t. This is possible since now it no longer affects the definiteness of Q. With this specification of Q he runs the Kalman Filter and estimates ψ by quasi maximum likelihood. Consistency and asymptotic normality are proved by [78]. In the following, the latter estimator will be referred to as the QML2 estimator.

In order to assess the small sample behavior of the QML estimators, Monte Carlo simulations have been conducted.[18] The Monte Carlo study by [78] provides evidence that QML2 beats QML1 both in terms of bias and variance.

A Markov Chain Monte Carlo (MCMC) approach is utilized by [52] for estimating one-, two- and three-factor versions of the CIR model. The MCMC

---

[18] See [40] and [78].

approach takes a Bayesian view, in which the model parameters ψ are also treated as random variables.[19] Given priors $p(\psi_i)$ for the elements of ψ, the aim of the procedure is to generate samples from the joint posterior $p(\alpha_1', \ldots, \alpha_T', \psi' | \mathcal{Y}_T)'$. Drawing from that posterior, smoothed states and estimates of the parameters can be constructed. [52] compare their Bayesian estimates with those obtained from classical QML approaches. It turns out that for some parameters the differences are substantial. Other versions of MCMC approaches for estimating term structure models from the affine class are employed by [103] and [93].

Next, we consider approaches that deal with the (additional) problem of nonlinearities in the state space model. For an exponential-affine model, for instance, nonlinearity arises if the measurement vector includes quantities which are not affine functions of yields.

Using an exponential-affine model with state-dependent variance and coupon bond prices as observable quantities leads to a state space model with nonlinear measurement equation and non-Gaussian transition equation. Estimation approaches for this type of state space model are discussed in [81], among them the "truncated second order filter.", which is is adopted by [8]. Without dwelling on technical details, the central inputs to the algorithm are Taylor approximations of the nonlinear measurement equation as well as the drift and diffusion term in the stochastic differential equations for the state. Parameter estimates are obtained by quasi maximum likelihood.

For a model with linear Gaussian transition equation and nonlinear measurement equation [79] suggests an "iterative extended Kalman Filter" (IEKF). In this framework, the prediction steps (5.20) and (5.21) remain unchanged, the one-step prediction for y_t is given by

$$\hat{y}_{t|t-1} = M(a_{t|t-1}).$$

The peculiar feature of the method is the way the updating is performed. [79] introduces it by interpreting Kalman filter updating as the solution to a weighted least squares problem:

$$\begin{aligned} a_t = &\operatorname*{argmin}_w (w - a_{t|t-1})' \Sigma_{t|t-1}^{-1} (w - a_{t|t-1}) \\ &+ (y_t - Mw - d)' H^{-1} (y_t - Mw - d). \end{aligned} \tag{8.28}$$

The updating for the nonlinear case is achieved by solving the same problem with general nonlinear measurement equation, i.e.

$$\begin{aligned} a_t = &\operatorname*{argmin}_w (w - a_{t|t-1})' \Sigma_{t|t-1}^{-1} (w - a_{t|t-1}) \\ &+ (y_t - M(w))' H^{-1} (y_t - M(w)). \end{aligned} \tag{8.29}$$

For a general nonlinear function M this does not have a closed form solution and [79] suggests to solve it using a Gauss-Newton optimization algorithm with the one-step prediction as starting value, i.e. $w^0 = a_{t|t-1}$.

[19] See [54] for MCMC estimation of state space models.

Filtering and parameter estimation with this method will in principle imply a considerable computational burden. For every iteration of the algorithm that performs the maximization of the quasi-likelihood, there are T GLS problems (8.29) that have to be solved numerically. The situation becomes even more complicated when for each iteration the gradient of the quasi-likelihood has to be computed numerically using finite differences. [79] circumvents the latter problem by developing an iterative algorithm that computes analytical derivatives for the quasi-likelihood.

In a Monte Carlo simulation the method is applied to the one-factor Vasiček model and the two-factor Beaglehole-Tenney model. For each model the measurement equation is based on coupon bond prices and additive normal errors. The parameters are estimated without bias and - except for the market prices of risk - with satisfying precision.

8.4 Model Adequacy and Interpretation of Results

After the estimation of a term structure model by one of the methods described above, it is important to assess the adequacy of the model specification. However, a systematic evaluation of estimation results is not always conducted as pointed out by [55]. Two types of questions are of interest. First, for a given model, are there indications that the model is misspecified? Second, if there are two or more candidate models, which of them should be preferred? We will start by considering approaches to answer the first type of question.

The first thing that is done in most studies is to check if the parameters are individually significant. Tests are mostly based on t-statistics, assuming asymptotic normality of parameter estimates. An alternative approach is taken by [16] who estimate confidence intervals for the parameters using the bootstrap. Furthermore, some of the parameters in the state space model have an economic interpretation and as such should have a certain dimension and sign. An important example are the parameters corresponding to the risk premia, see, e.g., [55].

In order to check for parameter stability over time, one can divide the sample in hand into subperiods and see if there is substantial difference between estimated parameters. This is done, e.g., by [66], [91], and [40].

The crucial feature of the term structure models discussed above is that they have been derived by imposing no-arbitrage conditions. As a consequence, the parameters in the measurement equation are functions of the parameters of the factor process and the risk premium parameters. [38] and [97] formally test on these restrictions using likelihood ratio tests. A Lagrange multiplier test for the model restrictions is applied by [40] for the case of a general exponential-affine model.

In a correctly specified model the size of the measurement error should not exceed a magnitude that can be justified by market imperfections and errors in measurement. Thus, the estimated standard deviation of ϵ in the measurement equation serves as an indicator for the misspecification of the model. The decision, however, which magnitude of the measurement error is acceptable is not based on a test. This criterion is frequently used in empirical studies, e.g. by [66], [39] or [9].

All of the exact and approximate filtering algorithms can generate one step ahead predictions of the observable y_t. If these are denoted by $\hat{y}_{t|t-1}$, the one-step-prediction errors (residuals) are defined as $v_t = y_t - \hat{y}_{t|t-1}$. In a correctly specified model, the standardized residuals $\tilde{v}_t$ should have mean zero and they should be serially uncorrelated. Some models imply in addition that $\tilde{v}_t \sim N(0, I)$. Thus, for the $\tilde{v}_t$, one can inspect the size of the mean, and test for autocorrelation and normality. Such a residual analysis is conducted by [55], [38] and [107]. Moreover, [55] point out that in a well-specified model the standardized residuals $\tilde{v}_t$ should not be correlated with the estimated factors. A violation of this property would indicate an omission of relevant factors.

A more informal assessment of model adequacy can be made by running a regression of yields of all maturities on a constant and the filtered or smoothed factors.[20] In a well specified model, the estimated regression coefficients should be similar to the corresponding estimated elements in the measurement equation. Such an investigation has been carried out by [38], [39] and [97].

As [55] point out, the state space approach imposes the parameter restrictions that come from the theoretical term structure model. This is different from standard factor analysis which is purely data-driven. When unconstrained factor processes are similar to the estimated state processes obtained from a model imposing theoretical restrictions, this can be interpreted as support for the model. Such similarities may be either checked by inspection of graphs or by computing a correlation coefficient.

Having the filtered or smoothed paths of the factors at hand one can use the estimated parameters of the measurement equation to construct a term structure for each time t of the sample. Summing for each yield over all t and dividing each sum by T, the number of observations, yields an average term structure as it is implied by the estimated model. This can then be compared to the observed average term structure as done by [38]. A similar comparison is made by [66] for the term structure of volatilities.

We now turn to the question of how to decide between two or more candidate models. In many cases, a comparison between two models is started by comparing the corresponding likelihood values. If the models are nested, one can test one model against the other using a likelihood ratio test as done, e.g., by [8], [97], [9] or [66].

More informally, [10] observe how the estimated standard deviations of the measurement errors decrease when moving to a model with one more factor.

[20] Such a regression may be performed in first differences when yields appear to be nonstationary.

[9] use filtered states to compute the implied yields $\hat{y}_{t|t}$.[21] These are plotted against the observed yields y_t. It is then judged by visual inspection whether the goodness of fit increases when using a two- instead of a one-factor model.

To distinguish between d and $d+1$ factor-models, information criteria are used by [107], [10], [9] and [37]. In order to discriminate between two non-nested models, [97] employs the J-test described in [37].

Using filtering and smoothing techniques it is possible to estimate the path of the unobservable state vector in a state space model. For exponential-affine multifactor models, many empirical studies find that the first factor can be interpreted as a level factor, whereas the second factor is often called a steepness factor. Third and higher order factors may have interpretations which are more subtle. For instance, [55] interprets the third factor as a kind of binary variable that increases the volatility of the very short end of the yield curve.

The latter described how to find out what the unobservable factors *do* rather then what they *are*. In general, the factor innovations are interpreted as economic news. Accordingly, the factors are interpreted as containing the weighted history of these innovations. However, sometimes their role is interpreted in more detail as, e.g., in the two-factor affine model by [91]. The first factor is interpreted as the real instantaneous interest rate, whereas the second is interpreted as expected inflation. Economic interpretations like that are typically based on some equilibrium derivation of a model. In order to check if such interpretations are empirically plausible one may proceed as follows. If there exists an observable proxy variable (e.g. the one-month treasury bill rate) that is assumed to correspond to the unobserved factor (e.g. interpreted as the instantaneous short rate), the measured evolution of the proxy variable and the estimated factor path can be compared graphically. If they move together quite well, this may be taken as support for the specific interpretation of the factor.

In the next chapter we will see some of the methods and techniques at work. It contains an empirical application of the material introduced so far. Three models from the AMGM class will be estimated using the state space model as the statistical framework.

[21] That is in the linear state space model $\hat{y}_{t|t} = M(\hat{\psi})a_t + d(\hat{\psi})$, where $M(\hat{\psi})$ and $d(\hat{\psi})$ denote system matrices based on estimated parameters and a_t is the filtered state.

9

An Empirical Application

In this chapter we present an empirical study in which we estimate three discrete-time models from the AMGM class. We use the data set of US treasury yields that has been presented in section 2.2. The main purpose of this chapter is to show the methodology - estimating term structure models in a state space framework - at work. Moreover, we want to point out what difference it can make to use a mixture model as opposed to a purely Gaussian model with the same number of factors.

9.1 Models and Estimation Approach

We estimate three models from the AMGM class: a Gaussian two-factor model, a two-factor model with a two-component mixture, and a three-factor Gaussian model. As introduced in chapter 3, the models are characterized by a vector-valued factor process[1]

$$X_t = \mathcal{K}X_{t-1} + u_t \tag{9.1}$$

and a specification of the stochastic discount factor (SDF), that is of the form

$$-\ln M_{t+1} = \delta + \iota' X_t + \lambda' u_{t+1}. \tag{9.2}$$

For the Gaussian models the factor innovation satisfies

$$u_t \sim N(0, V), \tag{9.3}$$

whereas for the mixture model

$$u_t \sim \sum_{b=1}^{B} \omega_b N(\mu_b, V_b), \quad \sum_{b=1}^{B} \omega_b = 1, \; \sum_{b=1}^{B} \omega_b \mu_b = 0. \tag{9.4}$$

[1] We repeat some of the formulas from chapter 3 for convenience. We also use the canonical representation of the models. That is why we have, for instance, no intercept in the equation of the factor evolution.

Going from the general to the specific, the factor process of the two-factor Gaussian model is given by

$$\begin{pmatrix} X_{1t} \\ X_{2t} \end{pmatrix} = \begin{pmatrix} \kappa_1 & 0 \\ 0 & \kappa_2 \end{pmatrix} \begin{pmatrix} X_{1\,t-1} \\ X_{2\,t-1} \end{pmatrix} + \begin{pmatrix} u_{1t} \\ u_{2t} \end{pmatrix} \tag{9.5}$$

where the distribution of the factor innovation is

$$\begin{pmatrix} u_{1t} \\ u_{2t} \end{pmatrix} \sim N\left(\begin{pmatrix} 0 \\ 0 \end{pmatrix}, \begin{pmatrix} v_1^2 & 0 \\ 0 & v_2^2 \end{pmatrix} \right). \tag{9.6}$$

The SDF satisfies

$$-\ln M_{t+1} = \delta + X_{1t} + X_{2t} + \lambda_1 u_{1\,t+1} + \lambda_2 u_{2\,t+1} \tag{9.7}$$

Interchanging the two factors will not alter the implied term structure.[2] For the Gaussian two-factor model and the mixture model that will be described hereafter, we will identify the factors by their persistence. That is, they are arranged such that $\kappa_1 > \kappa_2$.

Concerning the two-factor mixture model, the factor process and the SDF equation are of the same form as for the two-factor Gaussian model. The distribution of the factor innovation is specified as a Gaussian mixture with two components,

$$\begin{pmatrix} u_{1t} \\ u_{2t} \end{pmatrix} \sim \omega N\left(\begin{pmatrix} 0 \\ 0 \end{pmatrix}, \begin{pmatrix} v_{11}^2 & 0 \\ 0 & v_{21}^2 \end{pmatrix} \right) + (1-\omega) N\left(\begin{pmatrix} 0 \\ 0 \end{pmatrix}, \begin{pmatrix} v_{12}^2 & 0 \\ 0 & v_{22}^2 \end{pmatrix} \right) \tag{9.8}$$

We have tried three different specifications, one with $v_{11} \neq v_{12}$ and $v_{21} \neq v_{22}$, another with $v_{11} \neq v_{12}$ and $v_{21} = v_{22}$, and a third with $v_{11} = v_{12}$ and $v_{21} \neq v_{22}$. It turned out that the third one performed best and we only report the results of this specification. In order to identify the two components we assume that $v_{21} \geq v_{22}$. This assumption is embedded into the specification by parameterizing the first component variance as a multiple of the second. Summing up, we will assume that

$$v_{11} = v_{12} =: v_1, \quad \text{and} \quad v_{21}^2 = c_{22} v_{22}^2, \quad c_{22} \geq 1. \tag{9.9}$$

Finally, the three-factor Gaussian model consists of the factor process

$$\begin{pmatrix} X_{1t} \\ X_{2t} \\ X_{3t} \end{pmatrix} = \begin{pmatrix} \kappa_1 & 0 & 0 \\ 0 & \kappa_2 & 0 \\ 0 & 0 & \kappa_3 \end{pmatrix} \begin{pmatrix} X_{1\,t-1} \\ X_{2\,t-1} \\ X_{3\,t-1} \end{pmatrix} + \begin{pmatrix} u_{1t} \\ u_{2t} \\ u_{3t} \end{pmatrix} \tag{9.10}$$

with

[2] In the language of chapter 3, interchanging the factors corresponds to the invariant transformation with $l = \begin{pmatrix} 0 \\ 0 \end{pmatrix}$ and $L = \begin{pmatrix} 0 & 1 \\ 1 & 0 \end{pmatrix}$.

$$\begin{pmatrix} u_{1t} \\ u_{2t} \\ u_{3t} \end{pmatrix} \sim N\left(\begin{pmatrix} 0 \\ 0 \\ 0 \end{pmatrix}, \begin{pmatrix} v_1^2 & 0 & 0 \\ 0 & v_2^2 & 0 \\ 0 & 0 & v_3^2 \end{pmatrix}\right) \tag{9.11}$$

The pricing kernel is given by

$$-\ln M_{t+1} = \delta + X_{1t} + X_{2t} + X_{3t} + \lambda_1 u_{1\,t+1} + \lambda_2 u_{2\,t+1} + \lambda_3 u_{3\,t+1}. \tag{9.12}$$

Similarly as for the two-factor models, we assume that $\kappa_1 > \kappa_2 > \kappa_3$.

All three models have the property that both the matrix $\mathcal{K}$ and the (component) variance-covariance matrices are diagonal. For all three models, this implies that the factors are uncorrelated. Of course, this is a restrictive assumption and its validity could be tested for. For the two-factor models, correlation of the factors could be induced by introducing an additional free parameter for the (2,1)-element of $\mathcal{K}$. The hypothesis of uncorrelated factors would then correspond to this parameter being zero. Such a test, however, will not be conducted here and we will stick to the more simple specification.

Following from the no-arbitrage condition, all models imply that for each time to maturity n, the yield of a zero coupon bond has to satisfy

$$y_t^n = \frac{A_n}{n} + \frac{1}{n} B_n' X_t \tag{9.13}$$

where A_n and B_n are functions of the model parameters. They have to satisfy (3.72) - (3.73) for the Gaussian model and (3.80) - (3.81) for the mixture model.

The models are estimated making use of the methodology introduced in chapters 5 and 6. Each model is cast into its corresponding state space form and the parameters are estimated by maximum likelihood. For the Gaussian models, the state space model is linear and Gaussian, and the exact likelihood can be constructed using the Kalman filter. For the two-factor mixture model, the state space model is linear but the state innovations are distributed as a Gaussian mixture. For this model, we construct an approximate likelihood based on the AMF(1) filter.[3] We will now explain some details of the estimation process and turn to the results in the next section.

We use the data set presented in section 2.2. It contains measurements for $T = 444$ months. The data set contains time series of yields for maturities of 3, 6, 12, 24, 60, and 120 months. We use all of them for our estimations except the two-year yield. The latter will be spared for conducting certain diagnostic checks.

The yields are annualized, the models, however, hold for monthly yields. The theoretical models imply that for some arbitrary n, the joint evolution of factor and yield are given by (9.1) and (9.13). Then the annualized yield, $\tilde{y}_t^n := 1200 \cdot y_t^n$, satisfies[4]

[3] Using the AMF(2) filter delivered nearly the same results.

[4] We have to multiply by 1200 (and not by 12 only) since yields in the data set are expressed in percentages.

$$\tilde{y}_t^n = \frac{A_n^*}{n} + \frac{1}{n} B_n^{*\prime} X_t$$
$$X_t = \mathcal{K} X_{t-1} + u_t$$

with $A_n^* = 1200 \cdot A_n$ and $B_n^* = 1200 \cdot B_n$. It is this kind of representation that we use in the empirical study. It implies that the parameters that we obtain are those that correspond to the original monthly yields. Accordingly, they can be compared in size with parameters from the literature that have been obtained for other samples using possibly different statistical techniques. The reason for using annualized yields (as opposed to monthly yields) lies in the fact that for monthly yields the measurement error in the corresponding state space model would have a very low standard deviation (of around 7e-6). This would possibly lead to numerical difficulties.

We do not want to carry on with the tilde on top of our annualized yields, so we drop it from here on and understand each y_t^n as an annualized yield.

For the state space models associated with our theoretical term structure models, the measurement vector y_t is five-dimensional,

$$y_t = (y_t^{n_1}, y_t^{n_2}, \ldots, y_t^{n_5})', \quad (n_1, n_2, \ldots, n_5)' = (3, 6, 12, 60, 120)'.$$

For each term structure model we identify the factor vector with the state vector, i.e. $\alpha_t = X_t$. The measurement equation has the form

$$\begin{pmatrix} y_t^{n_1} \\ \vdots \\ y_t^{n_5} \end{pmatrix} = \begin{pmatrix} 1200 \cdot \frac{1}{n_1} A_{n_1} \\ \vdots \\ 1200 \cdot \frac{1}{n_5} A_{n_5} \end{pmatrix} + \begin{pmatrix} 1200 \cdot \frac{1}{n_1} B'_{n_1} \\ \vdots \\ 1200 \cdot \frac{1}{n_5} B'_{n_5} \end{pmatrix} \alpha_t + \begin{pmatrix} \epsilon_{1t} \\ \vdots \\ \epsilon_{5t} \end{pmatrix}. \tag{9.14}$$

where the functional forms of the A_{n_i} and the B_{n_i} differ across models, of course. Written more compact in the familiar notation of a state space model,

$$y_t = d + M\alpha_t + \epsilon_t. \tag{9.15}$$

For the measurement error we use the simple specification

$$\epsilon_t \sim N(0, h^2 I_5). \tag{9.16}$$

This is not an innocuous assumption since it implies that the difference between theoretical and observed yields has the same variance for all maturities. We also tried a specification in which the variances were allowed to be pairwise different. However, it turned out that the other parameter estimates have not been affected much by this change of specification.

The two- or three-dimensional state vector is governed by the transition equation

$$\alpha_t = T\alpha_{t-1} + \eta_t. \tag{9.17}$$

As outlined in chapter 8, the transition equation *is* in fact the evolution equation of the factor vector. The new notation is introduced in order to emphasize that we are now considering the statistical model, on which our estimation will be based. So we have $\alpha_t = X_t$, $T = \mathcal{K}$ and $\eta_t = u_t$. For the distribution of the state innovation we have for the two Gaussian models

$$\eta_t \sim N(0, Q) \tag{9.18}$$

and for the two-factor mixture model

$$\eta_t \sim \omega N(0, Q_1) + (1 - \omega) N(0, Q_2), \tag{9.19}$$

where $Q = V$ and $Q_i = V_i$.

For the two-factor Gaussian model, the unknown model parameters to be estimated are κ_1, v_1^2, λ_1, κ_2, v_2^2, λ_2, δ and h^2. The parameters κ_1, v_1^2, κ_2, v_2^2 of the theoretical model appear in both, the transition equation and the measurement equation, whereas the parameters λ_1, λ_2 and δ appear in the intercept vector d of the measurement equation only.

Concerning the four parameters v_1^2, λ_1, v_2^2 and λ_2, the model may be equivalently parameterized in v_1, $\lambda_1 v_1$, v_2 and $\lambda_2 v_2$.[5] This can be seen as follows. The only places in which the parameters λ_1 and λ_2 appear are the functions A_n. For a Gaussian model, according to (3.73), A_n is computed as

$$A_n = \sum_{i=0}^{n-1} G(B_i) \tag{9.20}$$

where

$$G(B_i) = \delta + B_i' a - \frac{1}{2}(\lambda + B_i)' V (\lambda + B_i).$$

With a diagonal V matrix, expanding the expression $(\lambda + B_i)' V (\lambda + B_i)$ yields

$$(\lambda + B_i)' V (\lambda + B_i) = \sum_{j=1}^{2} \lambda_j^2 v_j^2 + 2 B_{ij} \lambda_j v_j^2 + B_{ij}^2 v_j^2 \tag{9.21}$$

$$= \sum_{j=1}^{2} (\lambda_j v_j)^2 + 2 B_{ij} (\lambda_j v_j) \cdot v_j + B_{ij}^2 v_j^2 \tag{9.22}$$

where B_{ij}, $j = 1, 2$ denotes the jth component of B_i. Thus, λ_j only shows up as a multiplier of v_j.

The same argument goes through for the three-factor model, which will be parameterized in v_1, $\lambda_1 v_1$, v_2, $\lambda_2 v_2$, v_3, $\lambda_3 v_3$. A similar reasoning holds for the two-factor mixture model. For each mixture component b one can expand the exponent $(\lambda + B_i)' V_b (\lambda + B_i)$ in (3.81) in the same fashion as just shown for the Gaussian case.[6] Thus, our two-factor model is parameterized in v_1, $\lambda_1 v_1$, v_{22}, $v_{21} = (\sqrt{c_{22}} v_{22})$, and $\lambda_2 v_{22}$.

[5] This is also done by [28].

[6] The parameterization that we use would not be possible if $\mu_b \neq 0$ as can be seen from (3.81).

Estimating the model, it turned out that δ and the market price of risk parameters $\lambda_1 v_1$ and $\lambda_2 v_2$ cannot be estimated very accurately. Moreover, the estimated variance-covariance matrix shows that they are highly correlated.[7] In particular, the parameter $\lambda_1 v_1$ has been individually insignificant, so we dropped it from the model.

Summing up, the following parameters will be estimated. For the Gaussian two-factor model,

$$\kappa_1,\ v_1,\ \kappa_2,\ v_2,\ \lambda_2 v_2,\ \delta,\ h^2,$$

for the Gaussian two-factor mixture model,

$$\kappa_1,\ v_1,\ \kappa_2,\ v_{22},\ \lambda_2 v_{22},\ c_{22},\ \omega,\ \delta,\ h^2$$

and for the Gaussian three-factor model,

$$\kappa_1,\ v_1,\ \kappa_2,\ v_2,\ \lambda_2 v_2,\ \kappa_3,\ v_3,\ \lambda_3 v_3,\ \delta,\ h^2.$$

Note that some of the parameters have to satisfy certain restrictions. We have:

$$\begin{aligned} -1 \leq \kappa_i \leq 1,\ i = 1,2,3 &\quad \text{(stationarity of the factor process)} \\ v_i \geq 0,\ i = 1,2,3,\ \text{and}\, v_{22} \geq 0 &\quad (v_i \text{ and } v_{22} \text{ are standard deviations}) \\ c_{22} \geq 1 &\quad \text{(by our assumption above)} \\ 0 < \omega < 1 &\quad (\omega \text{ is a component weight}) \\ h^2 \geq 0 &\quad (h^2 \text{ is a variance}) \end{aligned}$$

One strategy for dealing with such restrictions is to simply ignore them: if unconstrained maximization of the likelihood leads to maximizers that satisfy the restrictions, everything will be fine. If, in addition, the estimates lie in the interior as opposed to on the border of the admissible parameter space, one can also rely on the usual asymptotics in order to compute approximate standard deviations. However, in practice, there is the additional problem that parameters may fall outside the admissible set when running a numerical algorithm for maximizing the likelihood. If, for example, the nth step of some gradient algorithm suggests a v_1 that is negative or too close too zero, the algorithm breaks down: with such a trial value for v_1 it may be impossible to compute a valid likelihood.

Therefore, we have reparameterized the likelihood along the lines of [57]. For instance, the standard deviation parameter v_1 in the two-factor models has been rewritten as a function of an auxiliary parameter ψ_2^{aux} as

$$v_1 = \left(\frac{\psi_2^{aux}}{100} \right)^2 .$$

[7] All of these three parameters only enter the intercept vector d and do not show up elsewhere in the model. However, there is no identification problem as one might suspect. All of these parameters are individually identified, since we use five yields in the measurement vector.

The squaring insures nonnegativity of v_1 while leaving the new parameter ψ_2^{aux} unconstrained. As for another example, the variance multiplier in the two-factor mixture model, is reparameterized as

$$c_{22} = 1 + (\psi_6^{aux})^2,$$

insuring that c_{22} is bigger than one. Similar reparameterizations have been performed for the other model parameters as well.

The reparameterizations also serve another purpose, namely that of having all auxiliary parameters ψ_i of similar magnitude. As evident from the results that will be shown below, the estimates of the original parameters can differ by a factor of up to 10^5 from each other.[8] This may cause a problem for optimization algorithms that have to compute numerical gradients of the likelihood function. Accordingly, our reparameterization is done in a way that all auxiliary parameters ψ_i^{aux} are in a range of approximately 0.4 to 14.

Denoting the vector of original parameters by ψ,[9] and the vector of auxiliary parameters by ψ^{aux}, the two are connected by an invertible function $p : \mathbb{R}^w \rightarrow \mathbb{R}^w$, $\psi = p(\psi^{aux})$, where w is the dimension of ψ^{aux} and ψ. The log-likelihood is written as a function of the new parameters, $\ln \mathcal{L}(\psi) = \ln \mathcal{L}(p(\psi^{aux}))$. The likelihood is maximized with respect to the auxiliary parameters ψ^{aux} yielding the maximizer $\hat{\psi}^{aux}$. By the invariance property of maximum likelihood estimators, the ML estimator for the original parameters is given by $\hat{\psi} = p(\hat{\psi}^{aux})$.

The Hessian of $\ln \mathcal{L}(p(\psi^{aux}))$ with the derivatives taken with respect to ψ^{aux} and evaluated at $\hat{\psi}^{aux}$ can be used to construct the estimated variance-covariance matrix of $\hat{\psi}^{aux}$ using (5.43). However, we are interested in the variance-covariance matrix of the estimates of the original parameters $\hat{\psi}$. Accordingly, we compute the matrix of second derivatives of $\ln \mathcal{L}(\psi)$ with respect to ψ, evaluated at $\hat{\psi}$. Its negative inverse divided by the number of observations then delivers the estimated variance-covariance matrix of our parameter estimates.

Before we come to the estimation results, some last technical comments concerning the estimation are in order.

The empirical study is conducted using GAUSS. Maximization of the likelihood is performed using the BFGS algorithm as implemented in GAUSS' MAXLIK package. Gradients of the likelihood are computed numerically. For each of the three models, the filters employed for constructing the (approximate) likelihood are initialized by their unconditional moments as given by (5.27) and (5.28). Starting values for the auxiliary parameters have been found by trial and error. In order to prevent reporting local maxima, we have tried different sets of starting values for each model. We have not come across an instant, in which two different starting values led to different estimates.

---

[8] For the mixture model, we have $\hat{v}_1 = 0.000263$ while $\hat{c}_{22} = 26.1$.

[9] That is, for instance, for the two-factor Gaussian model $\psi = (\kappa_1, v_1, \kappa_2, v_2, \lambda_2 v_2, \delta, h^2)'$.

9.2 Estimation Results

Table 9.1 contains the maximum likelihood estimates of the parameters. Estimated standard errors are given in parentheses. They are equal to the square root of the diagonal elements of the estimated variance-covariance matrix, computed according to (5.43).

Table 9.1. Estimation results for data set of US treasury yields

	Two Factors, Gaussian	Two Factors, Mixture	Three Factors, Gaussian
κ_1	0.998 (2.36e-4)	0.998 (2.38e-4)	0.999 (1.48e-4)
v_1	0.000285 (8.27e-6)	0.000263 (1.14e-5)	0.000281 (1.13e-5)
κ_2	0.949 (1.20e-3)	0.950 (1.94e-3)	0.954 (1.34e-3)
v_2	0.000439 (9.54e-6)		0.000510 (2.13e-5)
$\lambda_2 v_2$	-0.142 (0.0560)		-0.0465 (0.0442)
δ	0.0167 (6.19e-4)	0.0121 (2.53e-3)	0.0117 (9.77e-4)
v_{22}		0.000230 (2.10e-5)	
$\lambda_2 v_{22}$		-0.049 (0.0158)	
c_{22}		26.10 (8.130)	
ω		0.138 (0.0386)	
κ_3			0.687 (0.0150)
v_3			0.000506 (2.20e-5)
$\lambda_3 v_3$			-0.340 (0.0518)
h^2	0.0346 (6.96e-4)	0.0345 (1.28e-3)	0.00855 (3.95e-4)
$\ln \mathcal{L}(\hat{\psi})$	-479.90	-404.07	157.36
AIC	973.79	826.13	-294.72

Estimated standard errors are given in parentheses.

First of all, the dimension and sign of the estimates are reasonable for all parameters.

The first factor is highly persistent as the estimate of κ_1 is nearly one for all models. Estimated standard errors may be interpreted with some caution since the estimate is very close to the boundary of the parameter space. For future studies we suggest using the bootstrap in order to obtain reliable confidence intervals. The standard deviation v_1 of the first factor is estimated with satisfactory precision and it does not differ much across models.

The second factor exhibits lower autocorrelation (κ_2) than the first factor, but it is still very high. The innovation of the second factor is the place in which the Gaussian models differ from the mixture model. For the latter model, the marginal distribution of the factor innovation is a mixture of two normals,

$$u_{2t} \sim \omega N(0, v_{21}^2) + (1-\omega)N(0, v_{22}^2), \quad \text{with} \quad v_{21}^2 = c_{22} \cdot v_{22}^2.$$

Judging on the basis of a usual t-test, the estimate of the weight ω is significantly different from zero and the estimate of the variance ratio c_{22} is different from unity.[10] So the results suggest that for the sample at hand the density for the second factor innovation is in fact a 'true' mixture of normals. It can be interpreted in such a way that in 86.2 percent of the time the innovation is drawn from a normal with standard deviation $v_{22} = 0.00023$, and in 13.8 percent it is drawn from a normal whose standard deviation is $5.11 (= \sqrt{26.1})$ times larger.

For the mixture model, the estimates of v_{22}, ω and c_{22} imply that the estimate of the standard deviation of the second factor innovation is given by

$$\hat{v}_2 := \left(\hat{\omega} \cdot \hat{c}_{22} \cdot \hat{v}_{22}^2 + (1-\hat{\omega}) \cdot \hat{v}_{22}^2\right)^{0.5} = 0.000486.$$

This does not deviate much from the estimated standard deviation of the second factor innovation for the Gaussian two-factor model.

In the mixture model, the parameter estimates imply for the excess kurtosis of u_{2t},

$$\widehat{kurt}(u_{2t}) = \frac{3\left[\hat{\omega} \cdot \left(\hat{c}_{22} \cdot \hat{v}_{22}^2\right)^2 + (1-\hat{\omega}) \cdot \hat{v}_{22}^4\right]}{\hat{v}_2^4} - 3 = 11.284,$$

Recall that the excess kurtosis is zero (by definition) for the Gaussian models.

The two panels in figure 9.1 show the marginal densities of the factor innovations that are implied by the parameter estimates. The left panel contains the estimated densities of u_{1t}, the innovation of the first factor. The solid line corresponds to the Gaussian two-factor model, the dashed line corresponds to the mixture model. Recall that both densities are normal. They differ from each other due to the fact that they have slightly different variances. The right panel shows a more substantial difference. The solid line depicts the density of

---

[10] In face of the fact that we use the approximate likelihood generated by the AMF, the estimated standard deviations should be used with caution.

u_{2t} for the Gaussian model. The dashed line represents the density of u_{2t} for the mixture model. The density is a Gaussian mixture with two components. It is remarkably different compared to its Gaussian counterpart with which it shares nearly the same variance.

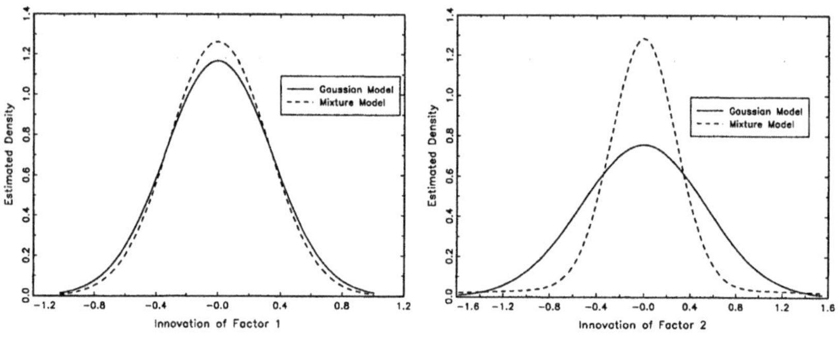

Fig. 9.1. For the two-factor models: Estimated densities of the innovation of the first factor (left panel) and the second factor (right panel)

For all three models, the market price of risk parameters, $\lambda_2 v_2$ and $\lambda_2 v_{22}$ have the expected negative sign which corresponds to a positive term premium. These parameters are estimated with lower relative precision compared to the other parameters discussed so far. For the three-factor model $\lambda_2 v_2$ is not even significantly different from zero. The parameter δ that governs the average level of the yield curve is individually estimated quite precisely. However, the estimated autocorrelation matrix of estimates (not reported here) shows that for all models considered, the correlation of the market price of risk parameters and δ is high.

Heuristically, these properties may be explained by the fact that the market price of risk parameters and δ only show up in the intercept vector d of the measurement equation. Since the factors have mean zero, it is easy to see from equation (9.15) that the vector d contains the individual means of yields included in y_t. Now, since all yields are highly autocorrelated, their means - and in turn the parameters that they depend on - cannot be estimated very precisely.[11]

For the three-factor model, the parameters κ_3, v_3, and $\lambda_3 v_3$ of the additional factor process had to be estimated. The estimate of the autocorrelation parameter κ_3 is remarkably smaller than the autocorrelation parameters of the first two factors. The estimated innovation variance ${v_3}^2$ is similar in size to that of the second factor. Unlike in the case of the second factor, the market price of risk parameter $\lambda_3 v_3$ is individually significantly different from zero.

[11] Compare a similar remark in [6].

The estimated variance $\hat{h}^2$ of the measurement error has the same size for both two-factor models. Recall that the measurement error captures the difference between observed annualized yields and the theoretical yields implied by the respective model under consideration. The estimates for the two-factor models imply that this error has a standard deviation of $0.186 (= \sqrt{0.0346})$ percentage points. The standard deviation implied by the three-factor model is half as large, it amounts to 0.092 percentage points.

The bottom of table 9.1 contains the values of the log-likelihood at maximum for the three models. We also provide the value of Akaike's information criterion, defined as

$$AIC = -2 \ln \mathcal{L}(\hat{\psi}) + 2w$$

where w is the number of unknown parameters. The AIC decreases in the value of the likelihood and increases in the number of parameters that have to be estimated. Using the AIC as a model selection criterion, the model with the smallest value of the AIC is chosen. Employing this measure for selecting one of our three models, the three-factor model would be preferred. Comparing between the two two-factor models only, the mixture model would beat the pure Gaussian model. A worthwhile exercise for future research would consist of choosing a mixture distribution for the innovations of the three-factor model and checking if this enhanced three-factor model beats the pure Gaussian one considered here.

Having discussed the content of table 9.1, we continue with some further analysis and interpretation of the results.

Figure 9.2 displays the average observed yield curve together with the average estimated yield curves for the three models.[12] For convenience the points of the average observed yield curve are connected. For $(n_1, n_2, \ldots, n_6)' = (3, 6, 12, 24, 60, 120)'$, the observed average yield curve consists of the points $(n_i, \bar{y}^{n_i})$, where

$$\bar{y}^{n_i} = \frac{1}{T} \sum_{t=1}^{T} y_t^{n_i}, \quad i = 1, \ldots, 6,$$

is the average of the annualized n_i-month yields over the 444 observations. Note that the 24-month yield, that has not been used for the estimation, is also included. The average estimated yield curve is given by the points $(n_i, \hat{y}^{n_i})$ where

$$\hat{y}^{n_i} - \frac{1}{T} \sum_{t=1}^{T} \left(\frac{\hat{A}_{n_i}}{n_i} + \frac{1}{n_i} \hat{B}'_{n_i} a_{t|t} \right). \tag{9.23}$$

Here $\hat{A}_{n_i}$ and $\hat{B}_{n_i}$ are the coefficient functions implied by the models where the parameters are replaced by their maximum likelihood estimates. The $a_{t|t}$ are the filtered states at time t. Thus, for a given time t, $a_{t|t}$ is an estimate

[12] The points representing the two-factor Gaussian model and those representing the two-factor mixture model nearly coincide and are hard to distinguish from each other.

of the factor vector X_t, which is constructed using all information up to this point in time. The figure shows that the mean yield curve is matched well by

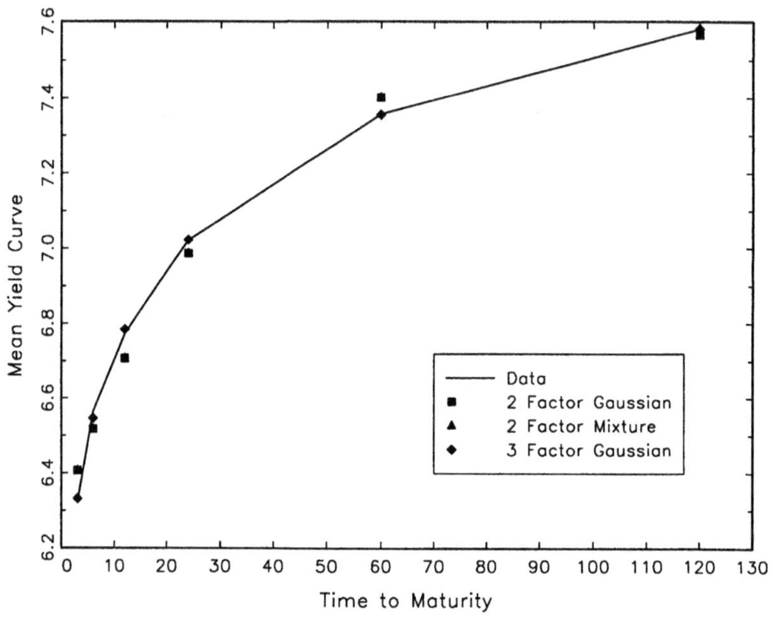

Fig. 9.2. Mean yield curve: observed and implied by estimated models

all models whereas the three-factor model seems to have a slight edge over the other two models.

In univariate time series analysis, diagnostic tests for fitted models are often based on residuals. In particular, residuals should be uncorrelated over time. Tests for the correlation of residuals are based on the autocorrelations of the estimated residuals. In multivariate time series analysis there is more than one autocorrelation for a given lag. Let $\{v_t\}$ be a vector valued series of residuals, where the v_t are each of dimension $N \times 1$. Then for a given lag k there are N^2 possibly different autocorrelations, namely between $v_{i\,t}$ and $v_{j,\,t-k}$ for all pairs (i, j), $i = 1, \ldots, N$, $j = 1, \ldots, N$.[13] In the literature that deals with the estimation of term structure models in a state space framework, the analysis is generally restricted to univariate autocorrelations, i.e those between $v_{i\,t}$ and $v_{i\,t-k}$.

For our models we want to provide two measures for the autocorrelation of residuals. First, we will show the five univariate autocorrelation functions.

[13] Note that in general the autocorrelation between $v_{i\,t}$ and $v_{j,\,t-k}$ is different from that between $v_{j\,t}$ and $v_{i,\,t-k}$.

Second, we provide a measure that tries to capture multivariate autocorrelation in a condensed form. We do not seek to formally test on autocorrelation of residuals by, for instance, using a multivariate portmanteau statistic. This is partly due to the fact that we do not know how such a statistic would behave for our model with Gaussian mixture innovations.

Following the notation in chapter 5, the residual vector v_t at time t is given by

$$v_t = y_t - y_{t|t-1},$$

where $y_{t|t-1}$ is the one-step forecast of y_t based on observations up to time $t-1$. The (i,j)-element of the autocorrelation matrix of v_t for lag k, $\Gamma(k)$, is given by[14]

$$\Gamma(k)_{ij} = \frac{\sum_{t=15+k}^{T}(v_{i\,t} - \bar{v}_i)(v_{j,\,t-k} - \bar{v}_j)}{\sqrt{\sum_{t=15+k}^{T}(v_{i\,t} - \bar{v}_i)^2} \cdot \sqrt{\sum_{t=15+k}^{T}(v_{j\,t} - \bar{v}_j)^2}}, \tag{9.24}$$

where

$$\bar{v}_l = \frac{1}{T-15}\sum_{t=16}^{T} v_{l\,t\cdot}, \quad l = i, j.$$

The first five panels in figure 9.3 depict the univariate autocorrelation functions $\Gamma(k)_{ii}$ for $i = 1, 2, \ldots, 5$. The graph in the lower right corner is intended to give an overall measure of autocorrelation of the residuals. For each lag k, we plotted for each model the norm $||\Gamma(k)||$ of the autocorrelation matrix $\Gamma(k)$ against the lag k. We have defined this norm as[15]

$$||\Gamma(k)|| := \max_{i,j}|\Gamma(k)_{i,j}|. \tag{9.25}$$

That is, for each lag k, the figure contains the largest (in absolute value) element of the 25 elements of the autocorrelation matrix.

The first thing to be noted is that the five univariate autocorrelation functions do not differ much across models. At lag 1 the ACFs assume their maximum with $\Gamma(1)_{ii}$ amounting to a level between about 0.2 and 0.3. For higher lags, the ACFs fluctuate around zero. Using the popular bounds of $\pm 2/\sqrt{T}$ which corresponds to the interval $[-0.097, 0.097]$ here, it turns out that for $i = 1$ (residuals of three-month yields) eight of the estimated autocorrelations fall outside this interval. This is the case for all three models. For 60-month yields, only three of the autocorrelations fall outside the interval. Overall, the autocorrelations of residuals appear to be a little too high, but the observed patterns do not point towards strong misspecification.

In the following, we will compare the three models with respect to their prediction performance. All predictions will be made for observations the dates

[14] Note that we have dropped the first 15 observations in order to remove any dependence on the initialization of the filters.

[15] Of course, there are several alternatives to define the norm of a matrix.

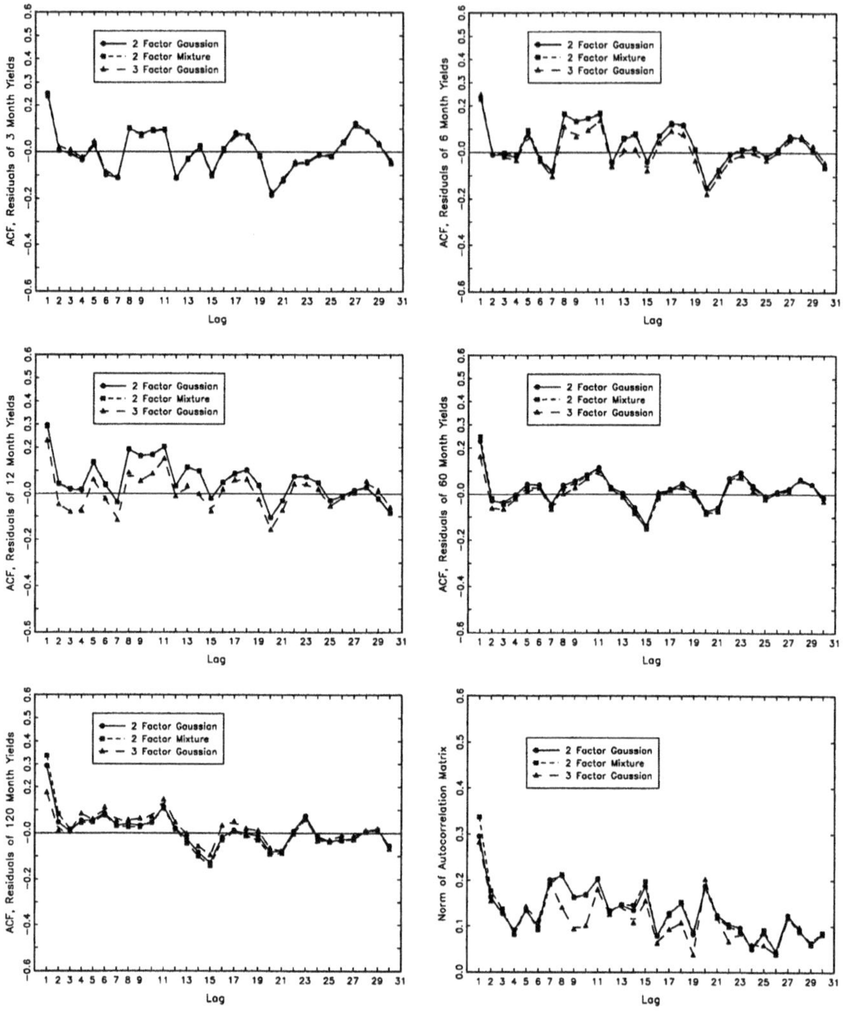

Fig. 9.3. First five panels (from left to right and top to bottom): ACF of the residuals of 3-, 6-, 12-, 60-, and 120-month yields. Right panel in the last row: norm of the multivariate autocorrelation matrices plotted against lags.

of which belong to our estimation period. If predictions are made for observations of 3-, 6-, 12-, 60- or 120-month yields, these are in-sample predictions since the time series of those yields have been used for parameter estimation. If predictions are made for the 24-month yield, such predictions are out-of-sample predictions. The term 'out-of-sample' refers to the cross-section, since the two-year yields are observed during our estimation period but have not been included in the set of data used for estimation.

Figure 9.4 shows the mean predicted yield curves together with the observed mean yield curve. This is essentially the same picture as figure 9.2 above, but the $\hat{y}^{n_i}$ are replaced by the average one-step predictions

$$\hat{y}^{n_i}_{pr} = \frac{1}{T}\sum_{t=1}^{T}\frac{\hat{A}_{n_i}}{n_i} + \frac{1}{n_i}\hat{B}'_{n_i}a_{t|t-1}. \tag{9.26}$$

The figure shows that on average, predicted yields match observed yields for

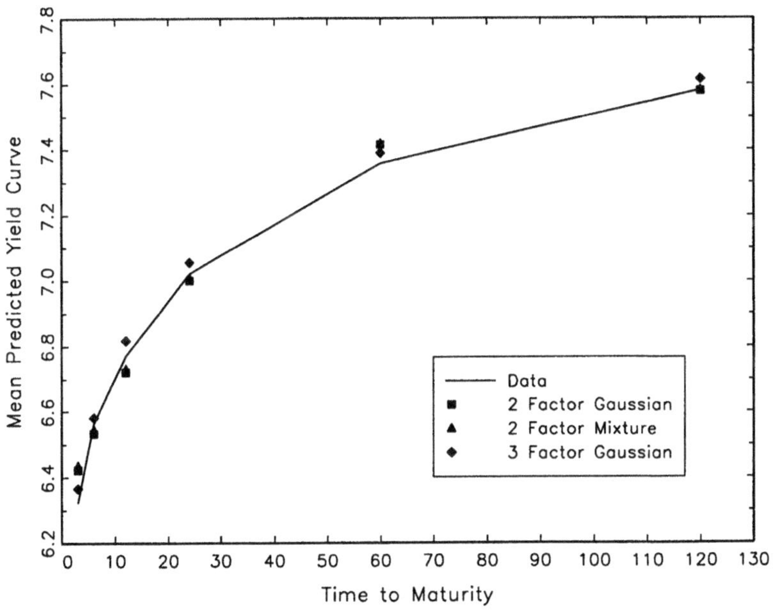

Fig. 9.4. Mean predicted yield curve

all models.

Table 9.2 gives the mean absolute errors for one-month predictions of yields. For the n_i-month yield the entry in the cell is given by

$$MAE(n_i) = \frac{1}{T-15}\sum_{t=16}^{T}\left|\hat{y}^{n_i}_{t|t-1} - y^{n_i}_t\right|. \tag{9.27}$$

Table 9.2. Mean absolute error for one-month prediction

Time to maturity	Two Factors, Gaussian	Two Factors, Mixture	Three Factors, Gaussian
3	0.346	0.349	0.353
6	0.366	0.367	0.362
12	0.400	0.399	0.369
60	0.301	0.302	0.292
120	0.263	0.267	0.246

For each of the three models, the prediction errors seem to rise with time to maturity first and then to fall again. While the two two-factor models perform similarly, the three-factor model shows a slightly superior performance except for the three-month yield.

Table 9.2 contains the mean absolute prediction errors for those maturities that have been employed for the estimation. As mentioned above, the 24-month yield, which is also in our data set, has not been used for estimation. However, the estimated model can be used to compute yields for this maturity as well. That is, we compute filtered estimates $\hat{y}_t^{24}$ and one-step predictions $\hat{y}_{t|t-1}^{24}$. The first type of estimate uses observations of the other five yields up to time t, the second type uses observations of the other five yields up to time $t-1$.[16]

Table 9.3. Mean absolute error for out-of-sample predictions

	Two Factors, Gaussian	Two Factors, Mixture	Three Factors, Gaussian
$\hat{y}_t^{24}$	0.163	0.164	0.094
$\hat{y}_{t\|t-1}^{24}$	0.373	0.371	0.347

Again, the three factor-model makes predictions that are a little better than those of the other two models. Especially the result from filtering seems satisfying. Having observed the other five yields until time t, the three-factor model's result for the two-year yield at time t, $\hat{y}_t^{24}$, deviates on average by less than 0.1 percentage points from the observed yield y_t^{24}.

One way to judge the *relative* forecasting performance of a model is to compare its forecasts with that resulting from a random walk assumption. As a measure for relative performance we use the ratio of the mean squared error resulting from the model's forecast and that from the naive forecast. It can be computed for the general case of s-step predictions. Denote by $\hat{y}_{t|t-s}$ the

[16] In other words, the first one uses the filtered state vector, whereas the second one uses the predicted state vector. Note however, that the estimates of the system parameters are in both cases based on the whole sample of 444 observations for 3-, 6-, 12-, 60- and 120-month yields but not for 24-month yields.

model's prediction for y_t based on observations up to time $t-s$.[17] Then, for s-step predictions, the MSE ratio $\gamma(s)$ for yields with maturity n_i is computed as[18]

$$\gamma(s) = \frac{\sum_{t=16}^{T} \left(\hat{y}_{t|t-s}^{n_i} - y_t^{n_i} \right)^2}{\sum_{t=16}^{T} \left(y_{t-s}^{n_i} - y_t^{n_i} \right)^2}. \tag{9.28}$$

If $\gamma(s) = 1$ the random walk forecast and the model forecast have the same quality, for $\gamma(s) < 1$ the model beats the random walk and vice versa. We have computed this measure for one-month and one-year forecasts, i.e. for $s = 1$ and $s = 12$. Table 9.4 shows the results.

Table 9.4. Ratio of the MSE from model prediction and the MSE from a random walk prediction

Time to maturity	Two Factors, Gaussian		Two Factors, Mixture		Three Factors, Gaussian	
	1 mon.	12 mon.	1 mon.	12 mon.	1 mon.	12 mon.
3	1.112	1.075	1.117	1.078	1.117	1.071
6	0.999	1.116	0.999	1.108	1.059	1.105
12	1.063	1.126	1.061	1.115	1.049	1.110
60	1.059	1.083	1.063	1.077	1.013	1.071
120	1.238	1.115	1.277	1.120	1.092	1.128

One-month prediction (left entry), twelve-month prediction (right entry)

For some affine multifactor models in continuous time, it is known that they have a rather poor forecasting performance.[19] The same result is obtained for the discrete-time models considered here. For the three models and for both forecast horizons, the model forecasts are inferior compared to the naive forecasts.[20] On the whole, the relative forecasting performance is worse for one-year predictions than for one-month predictions. Moreover, the three-factor model does a little better than the two-factor models.

Up to now we have said little about the factors that drive the term structure. Recall that with the filtering techniques presented in chapters 5 and 6 at hand we are able to estimate the path of the unobservable factors. Figure 9.5 depicts the estimated paths of the first and second factor for our two-factor

---

[17] Optimal multistep predictions for Gaussian and mixture models have been discussed in chapters 5 and 6, respectively.

[18] Note that again we started the summation at $t = 16$ only.

[19] See [42] who also provides the intuition behind this deficiency of affine models.

[20] However, forecasts generated by the models have an advantage compared to the random walk as they make a prediction that is consistent with the no-arbitrage restriction. That is, in forecasting the yield curve the model forecast imposes a cross-section restriction that the random walk does not obey.

models. That is, we have drawn the first and second component of the filtered state vector $a_{t|t}$ (multiplied by 1200) against time. The first thing to note is

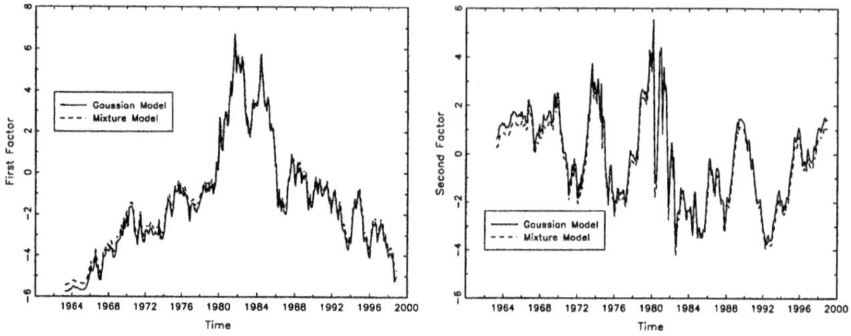

Fig. 9.5. Filtered process of the first factor (left panel) and the second factor (right panel)

that the results for the Gaussian model and the mixture model are similar. Second, comparing with figure 2.1 above, the path of the first factor seems to resemble the pattern of the evolution of the level of the yield curve.[21] In fact, the correlation between the filtered factor process and yields is high for each maturity. For both two-factor models, it reaches from 0.80 (correlation with the three-month yield) to 0.99 (correlation with the ten-year yield). Similar results are obtained for the three-factor model where the correlation is between 0.78 and 0.99. Against this background, the first factor may be referred to as a level factor.

This interpretation is supported if we look at the estimated factor loadings of the two-factor Gaussian model in figure 9.6.[22] The factor loading of the ith factor on the n-month yield is given by the ith component of the vector B_n/n in equation (9.13). Note that for all models considered, the vector B_n/n only depends on the κ_i parameters. In our models with diagonal $\mathcal{K}$ matrices, the ith component of B_n is simply given by κ_i^n. The interpretation of an arbitrary point on one of the curves of factor loadings is as follows: if that factor is increased, ceteris paribus, by one unit, the yield with time to maturity n is increased by the amount given on the axis of ordinates. Here, an increase in the first factor shifts up yields of all maturities nearly proportionally. Hence, the name level factor is justified. The second factor leads to a shift in the term structure that is strong at the short end of the yield curve and becomes weaker

[21] Of course it is parallel shifted by some amount, since the factor process has mean zero by assumption.

[22] The results of the two-factor mixture model imply nearly the same picture, therefore it is not shown here.

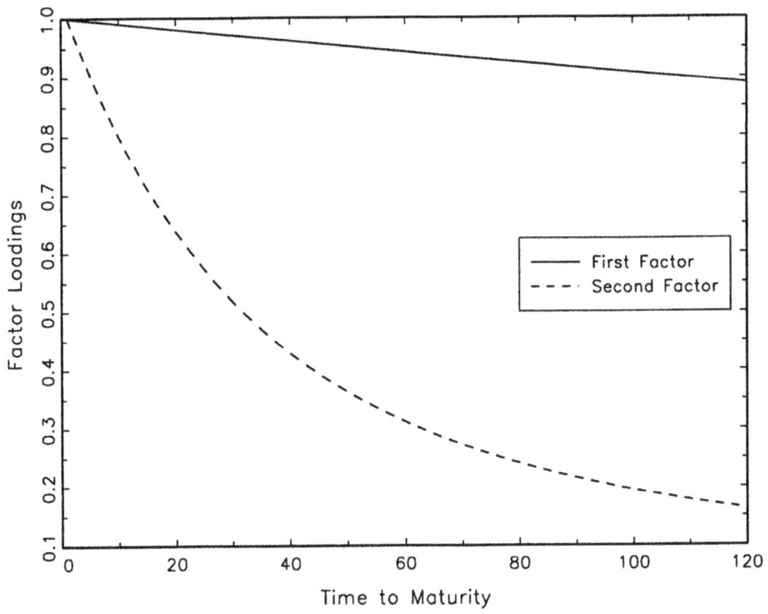

Fig. 9.6. Factor loadings for the two-factor model

as time to maturity rises. Accordingly, the second factor may be referred to as a twisting factor.

For the three-factor model, the same type of picture is drawn. Figure 9.7 shows that the first two factors can be given the same interpretation as before for the two-factor models. The additional factor works mostly at the short end of the yield curve.

Figure 9.8 shows the term structure of volatility, i.e. standard deviations of first differences in yields. The solid line connects the empirical standard deviations computed from the data. These are drawn together with the standard deviations that the estimated models imply for these yield changes. In the appendix, equation (E.30), we derive the variance for yield changes in our multifactor models. The values in the graph are computed according to the formulas in the appendix, where the parameters are replaced by the maximum likelihood estimates. The figure shows that all models imply a volatility curve that is decreasing in time to maturity. For maturities of two, five and ten years, the three-factor model comes closer to the observed volatility, but it overestimates the volatility at the short end. The two-factor models, in contrast, underestimate the volatility curve for maturities that exceed one year. However, all of these comparisons have to be made with caution since even for differenced yields we do not know how well the empirical standard deviations estimate the true ones.

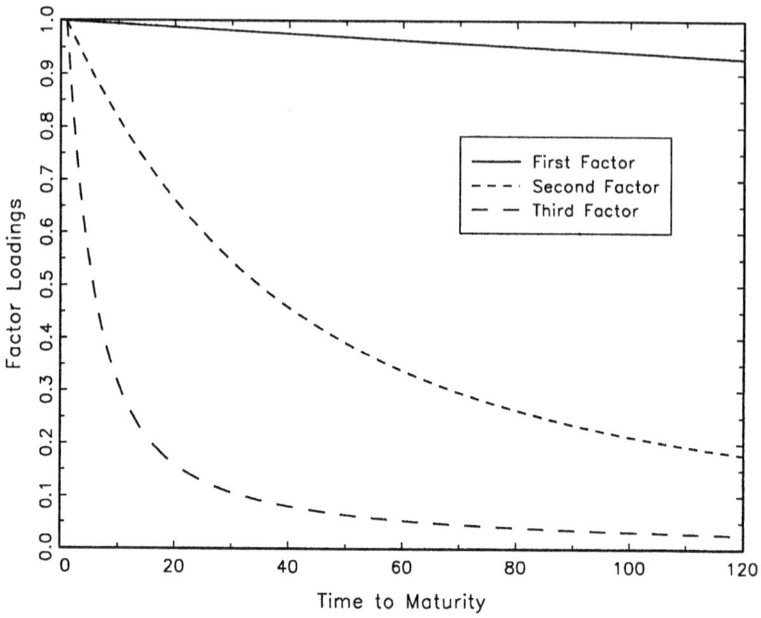

Fig. 9.7. Factor loadings for the three-factor model

Our last comments on the estimation results focus on the difference between the Gaussian models and the mixture model. As already pointed out, table 2.3 shows that yield changes exhibit considerable excess kurtosis. Multifactor term structure models with Gaussian innovations, however, imply zero excess kurtosis for yields in levels and yields in first differences. Formula (E.36) in the appendix gives the kurtosis of differenced yields implied by multifactor models with mixture innovations. Based on the maximum likelihood estimates of the parameters, the kurtosis has been computed for the maturities in the data set. These measures of kurtosis are graphed together with their empirical counterparts in figure 9.9. The important point to note is that the model in fact implies that the kurtosis is different from zero and that it decreases with maturity. Gaussian models imply a kurtosis which is identically zero for all maturities. One-factor models with mixture innovations, as discussed by [11] are capable of generating excess kurtosis, but the latter is constant for all maturities. Thus, concerning the matching of fourth moments, our simple two-factor model can be regarded as a step in the right direction.

Up to now we have discussed to what extent our three models are able to capture the behavior of first, second and fourth moments of yields in levels or first differences. Now, we want to look at the distribution as a whole. This will be done exemplarily for the three-month yield, representing the short end

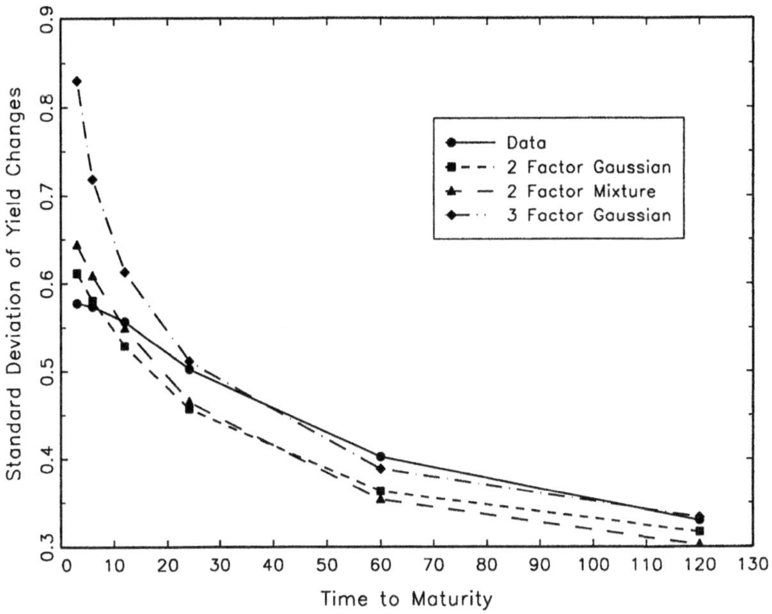

Fig. 9.8. Standard deviation of yield changes

of the yield curve, and the five-year yield, representing longer maturities. The analysis is done for first differences again.

The solid line in figure 9.10 depicts a kernel estimate for the distribution of Δy_t^3. It is based on our 444 observations and uses a Gaussian kernel with bandwidth $b = 1.364\hat{\sigma}N$, where N is the number of observations and $\hat{\sigma}$ is their standard deviation. The other lines are the density functions implied by the estimated models.

We derive the unconditional density of Δy_t^3 implied by the two-factor mixture model by means of a Monte Carlo simulation. Based on the maximum likelihood estimates of the parameters, we generate 10,000 observations of Δy_t^3 from the mixture model.[23] The corresponding kernel estimate of the density is shown in figure 9.10. In order to work under the same conditions for all models, the densities for the Gaussian models have been generated by analogous simulations.

The figure suggests that the two-factor mixture model captures the shape of the density best, followed by the two-factor Gaussian model. The density implied by the three-factor model does not appear to capture the distribution well.

We also use QQ-plots for comparing the distributions implied by the models with that given from the data. The QQ-plots in figure 9.11 (three of them

[23] The observations are generated without superimposing a measurement error.

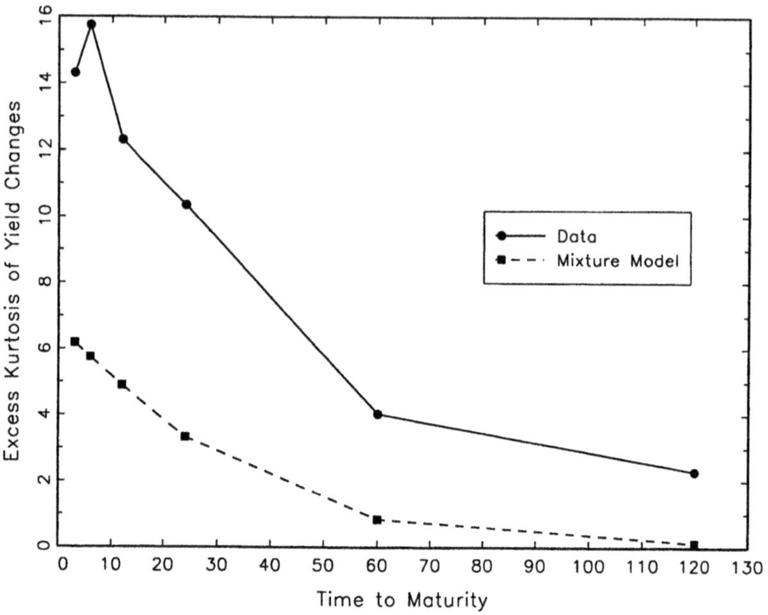

Fig. 9.9. Excess kurtosis of yield changes

drawn into one graph) are based on the probabilities 0.01, 0.02, ..., 0.99. For each of these probabilities, the corresponding quantile implied by the models is plotted against the empirical quantile of the data. If the points corresponding to a model were lying on the 45 degree line, this model would share the same quantiles with the data. Deviations from that line can be interpreted as a measure of distance between the two distributions. Like the density plot above, the QQ-plots suggest that the distribution implied by the two-factor mixture model comes closer to the distribution of the data than that implied by the two-factor Gaussian model. Again, the three-factor model performs worst.

A similar ranking can be inferred by looking at five-year yields. For changes of the five-year yield, figure 9.12 contains the three densities implied by the models as well as the density estimated from the data. The QQ-plots in figure 9.13 suggest that for the five-year yield, the advantage of the two-factor mixture model over the other two models shows up quite clearly.

9.3 Conclusion and Extensions

The empirical study has illustrated how the statistical state space model can be utilized for estimating dynamic term structure models. We have estimated

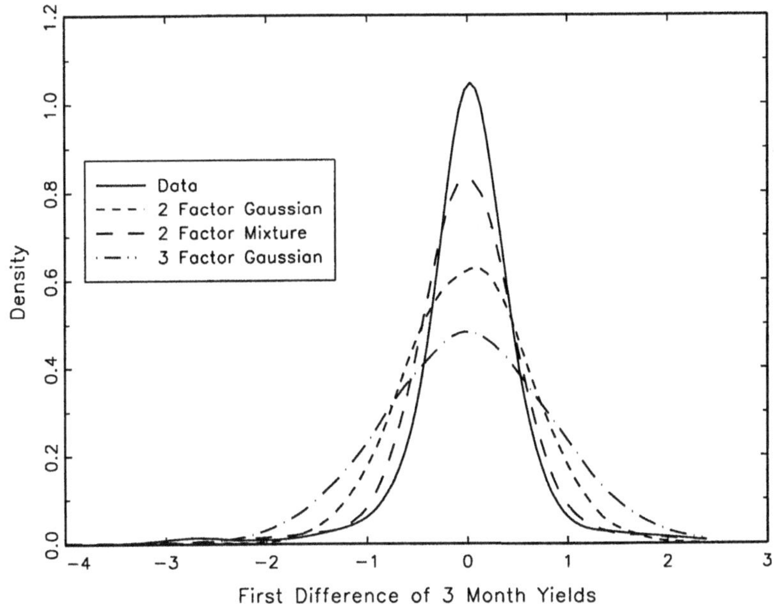

Fig. 9.10. Density of monthly changes in three-month yield

two Gaussian models and one model involving a Gaussian mixture distribution. For the Gaussian models, maximum likelihood estimation based on the Kalman filter has been conducted. For the mixture model, we have employed the AMF(1) which has been introduced in this book. It has turned out that estimation of the mixture term structure model is in fact feasible using this estimation device.

The parameter estimates are reasonable in size and have the correct signs. The autocorrelations of residuals do not point towards severe misspecification, but of course a more in-depth analysis should be conducted. The three-factor model is selected by the AIC and performs slightly superior than the other two with respect to some selected measures of fit. However, with respect to higher moments of the data, the two-factor mixture model appears to have an edge over the two Gaussian models.

For future research an integration of the mixture specification into the three-factor model is imaginable. Furthermore, a more elaborate specification of the mixture distribution could be tried. More reliable standard errors for the parameter estimates may be obtained by using the bootstrap. Moreover, a detailed analysis of the filtered factor paths is desirable. Although these factors are treated as latent within the model, one could conduct an ex-post exploration, relating the estimated paths to observable economic quantities.

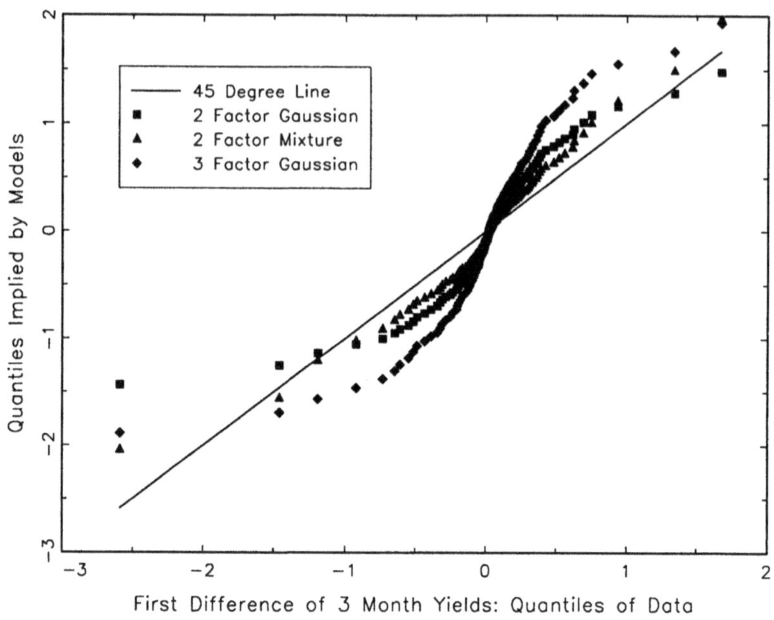

Fig. 9.11. QQ-plots for monthly changes in three-month yield.

With the state space approach it is in principle possible to include an arbitrary number of yields in the measurement vector. However, increasing that number leads to a higher computational burden. Thus, it is a sensible question which yields (i.e. corresponding to which maturities) should be included into the measurement vector. In a small Monte Carlo simulation (not reported here) we found out that, for a given dimension of the measurement vector, the selection of yields does have an impact on the precision of parameter estimates.

Finally, it would be instructive to use a different sample period and another country for the estimation. With the special focus of this book in mind, it would be particularly interesting to see how the estimated distributions of factor innovations change when the sample changes.

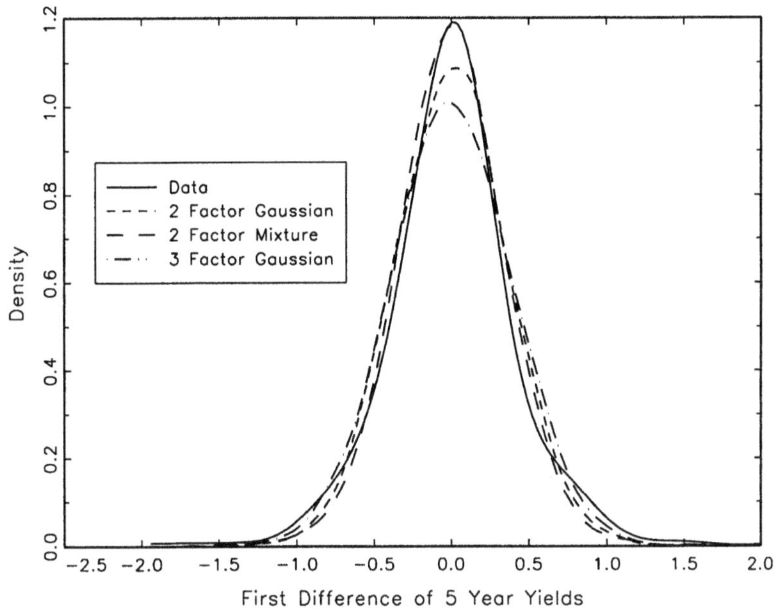

Fig. 9.12. Density of monthly changes in five-year yield

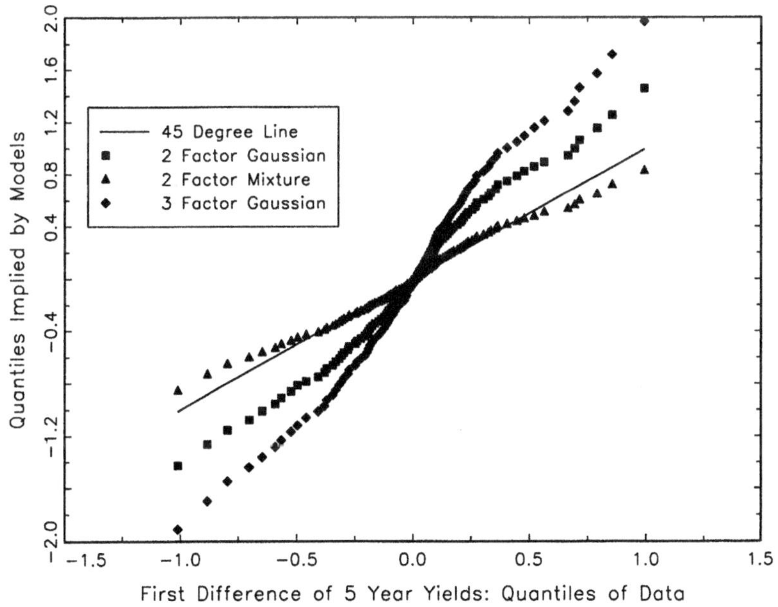

Fig. 9.13. QQ-plots for monthly changes in five-year yield.

10

Summary and Outlook

In this book we have dealt with the construction of term structure models and their estimation in a state space framework. We now summarize the central results and point out some directions for future research.

For the one-factor models in discrete time, various properties, in particular moments of yields and yield changes, have been derived. The one-factor models exhibit properties that cannot be reconciled with the stylized facts of term structure data. For instance, the models imply that yields are perfectly correlated in the cross-section and yields of all maturities have the same autocorrelation. The substantial excess kurtosis of bond yield changes observed in the data cannot be accounted for by the Gaussian model. The one-factor model suggested by [11], that replaces the simple normal distribution by a Gaussian mixture, is able to generate excess kurtosis, but the latter does not change with maturity.

As a generalization of the one-factor model by Backus et al., we have introduced d-factor models, for which the distribution of factor innovations is a Gaussian mixture with B components. The family of all these models has been termed the class of affine multifactor Gaussian mixture (AMGM) models. Models from the AMGM class have the convenient property that yields are affine in factors. A canonical representation for AMGM models has been proposed which is characterized by a parsimonious parameterization.

Another focus has been on state space models for which the state innovation is distributed as a Gaussian mixture. The exact filtering densities implied by those models are also Gaussian mixtures and could be computed exactly. However, the number of components in these mixtures increases exponentially with time, so the exact filter is not applicable for empirical time series that contain more than a few observations. To circumvent this problem, an approximate filter, called AMF(k), has been proposed. It can be employed for filtering and prediction. An approximate likelihood based on the AMF(k) can be employed for parameter estimation in the mixture state space model. The degree of precision can be regulated by the parameter k. Monte Carlo simulations based on three different data generating processes show results that are quite satisfactory.

In part three we have shown how the state space form can be used as a vehicle for estimating dynamic term structure models. It has been described how the theoretical model can be cast into state space form. Based on examples from the literature, the problems of estimation, diagnostics checking and model evaluation have been discussed. It has been argued that the state space model is a framework that is very suitable for the estimation of term structure models, making it possible to use time series and cross section information simultaneously.

In our own empirical study, three models from the AMGM class have been estimated using data on US treasury yields. We have used a two-factor and a three-factor Gaussian model and a two-factor mixture model. The two purely Gaussian models have been estimated by maximum likelihood based on the Kalman filter. For the two-factor mixture model the AMF(k) has been employed. Overall, the three-factor Gaussian model performs best. However, with respect to capturing the distribution of yield changes, the mixture model appears to have an edge over the other two models.

Concerning directions for future research, it may be explored how models from the AMGM class can be utilized for pricing interest rate contingent claims. As stated above, the intersection of the AMGM class and the exponential-affine class of [45] is the subclass of linear multifactor models with purely Gaussian innovations. It may be worthwhile to combine the structure of Duffie and Kan's models, in which volatility is level-dependent, with that of the AMGM models. A third idea is to make the mixture components depending on some additional explanatory variables that represent the business cycle. Such a specification may be interpreted as an alternative to recently proposed factor models with Markov regime switching.

Generally, there is room for exploring the properties of the AMF(k) further. For example, one may search for inequalities that characterize the relative precision of the AMF(k) compared to the Kalman filter or the exact filter. Second, we have seen that based on the AMF(k), an approximate likelihood can be constructed, which can be used for parameter estimation. Again, the properties of these estimators should be explored analytically. For obtaining standard errors for small samples, one should try using the bootstrap. Finally, additional Monte Carlo studies for the AMF(k) could be conducted, particularly in order to assess the relative performance compared to other approximate filters.

With respect to our empirical study, the very simple mixture specification of the two-factor model could be enhanced. Moreover, the mixture specification should also be tried within the three-factor model. The models should be estimated using data of different markets and samples periods. An application to corporate bonds (as opposed to government bonds) is also conceivable.

A

Properties of the Normal Distribution

Lemma A.1. *Let X be a g-dimensional random vector with a multivariate normal distribution, $X \sim N(\mu, Q)$, and let $c \neq 0$ be a g-dimensional vector of constants. Then*

$$E(e^{c'X}) = e^{(c'\mu + \frac{1}{2}c'Qc)} \tag{A.1}$$

Proof. The result follows directly from observing that the scalar $c'X$ has a univariate normal distribution $c'X \sim N(c'\mu, c'Qc)$. Thus, $e^{c'X}$ has a univariate lognormal distribution. □

Lemma A.2. *Let y and x be vectors of dimension g and k respectively and let A, a, C, b and D be vectors and matrices of appropriate dimension. Then the following equality holds:*

$$\begin{aligned} &\phi(y;\, Ax + a, C) \cdot \phi(x;\, b, D) \\ &= \phi\left(\begin{pmatrix} y \\ x \end{pmatrix}; \begin{pmatrix} Ab + a \\ b \end{pmatrix}, \begin{pmatrix} R_{yy} & R_{yx} \\ R_{xy} & R_{xx} \end{pmatrix}\right) \end{aligned} \tag{A.2}$$

where $R_{yy} = C + ADA'$, $R_{yx} = AD = R'_{xy}$, $R_{xx} = D$. In addition,

$$\int \phi(y;\, Ax + a, C) \cdot \phi(x;\, b, D) dx = \phi(y;\, Ab + a, R_{yy}) \tag{A.3}$$

Proof. The product of the two densities is

$$\begin{aligned} &\phi(y;\, Ax + a, C) \cdot \phi(x;\, b, D) \\ &= \frac{1}{(2\pi)^{g/2}\sqrt{|C|}} e^{-\frac{1}{2}(y - Ax - a)'C^{-1}(y - Ax - a)} \cdot \frac{1}{(2\pi)^{k/2}\sqrt{|D|}} e^{-\frac{1}{2}(x-b)'D^{-1}(x-b)} \end{aligned} \tag{A.4}$$

Introducing the transformation

$$\tilde{x} = x - b, \quad \tilde{y} = y - Ab - a,$$

allows the sum of the two exponents in (A.4) to be written as

$$
\begin{aligned}
&-\frac{1}{2}\left((\tilde{y}-A\tilde{x})'C^{-1}(\tilde{y}-A\tilde{x})+\tilde{x}'D^{-1}\tilde{x}\right)\\
&=-\frac{1}{2}\left(\begin{pmatrix}\tilde{y}-A\tilde{x}\\ \tilde{x}\end{pmatrix}'\begin{pmatrix}C & 0\\ 0 & D\end{pmatrix}^{-1}\begin{pmatrix}\tilde{y}-A\tilde{x}\\ \tilde{x}\end{pmatrix}\right)\\
&=-\frac{1}{2}\left(\begin{pmatrix}\tilde{y}\\ \tilde{x}\end{pmatrix}'\begin{pmatrix}I_g & -A\\ 0 & I_k\end{pmatrix}'\begin{pmatrix}C & 0\\ 0 & D\end{pmatrix}^{-1}\begin{pmatrix}I_g & -A\\ 0 & I_k\end{pmatrix}\begin{pmatrix}\tilde{y}\\ \tilde{x}\end{pmatrix}\right)\\
&=-\frac{1}{2}\left(\begin{pmatrix}\tilde{y}\\ \tilde{x}\end{pmatrix}'\begin{pmatrix}I_g & A\\ 0 & I_k\end{pmatrix}'^{-1}\begin{pmatrix}C & 0\\ 0 & D\end{pmatrix}^{-1}\begin{pmatrix}I_g & A\\ 0 & I_k\end{pmatrix}^{-1}\begin{pmatrix}\tilde{y}\\ \tilde{x}\end{pmatrix}\right)\\
&=-\frac{1}{2}\left(\begin{pmatrix}\tilde{y}\\ \tilde{x}\end{pmatrix}'\left[\begin{pmatrix}I_g & A\\ 0 & I_k\end{pmatrix}\begin{pmatrix}C & 0\\ 0 & D\end{pmatrix}\begin{pmatrix}I_g & A\\ 0 & I_k\end{pmatrix}'\right]^{-1}\begin{pmatrix}\tilde{y}\\ \tilde{x}\end{pmatrix}\right)\\
&=-\frac{1}{2}\left(\begin{pmatrix}\tilde{y}\\ \tilde{x}\end{pmatrix}'\begin{pmatrix}C+ADA' & AD\\ DA' & D\end{pmatrix}^{-1}\begin{pmatrix}\tilde{y}\\ \tilde{x}\end{pmatrix}\right)
\end{aligned}
$$

Summing up, we have

$$
\phi(y;\, Ax+a, C)\cdot\phi(x;\, b, D)=\frac{1}{(2\pi)^{\frac{g+k}{2}}\sqrt{|C||D|}}e^{-\frac{1}{2}\tilde{z}'R^{-1}\tilde{z}}, \tag{A.5}
$$

where

$$
R:=\begin{pmatrix}C+ADA' & AD\\ DA' & D\end{pmatrix},\quad \text{and } \tilde{z}:=\begin{pmatrix}\tilde{y}\\ \tilde{x}\end{pmatrix}.
$$

Using the result in [77, p. 50] concerning the determinant of partitioned matrices one obtains the equality

$$
|R|=|C||D|.
$$

Thus, the product of the two densities above can be written as a $g+k$-variate normal density. Retransforming $\tilde{z}$ back to the original $(y,x)'$ yields the proposed result:

$$
\begin{aligned}
&\phi(y;\, Ax+a, C)\cdot\phi(x;\, b, D)\\
&=\phi(\tilde{z};\, 0, R)\\
&=\phi\left(\begin{pmatrix}y\\ x\end{pmatrix};\begin{pmatrix}Ab+a\\ b\end{pmatrix},\begin{pmatrix}C+ADA' & AD\\ D'A' & D\end{pmatrix}\right).
\end{aligned}
$$

Taking the integral with respect to x leads to the marginal density of y. □

Lemma A.3. *Let x and y be random vectors such that the joint distribution of the vector $(x', y')'$ is the multivariate normal with mean and variance-covariance matrix given by*

$$\mu = \begin{bmatrix} \mu_x \\ \mu_y \end{bmatrix} \quad \textit{and} \quad C = \begin{bmatrix} C_{xx} & C_{xy} \\ C_{yx} & C_{yy} \end{bmatrix} \tag{A.6}$$

respectively. Then the distribution of x conditional on y is also multivariate normal with mean and variance-covariance matrix given by

$$\mu_{x|y} = \mu_x + C_{xy}C_{yy}^{-1}(y - \mu_y) \quad \textit{and} \quad C_{xx|y} = C_{xx} - C_{xy}C_{yy}^{-1}C_{yx} \tag{A.7}$$

Proof. See [59]. □

B

Higher Order Stationarity of a VAR(1)

We provide some results concerning stationarity up to fourth order for a VAR(1) process. These results are employed within the derivation of moments for the discrete-time term structure models.

Proposition B.1. *Let $\{X_t\}$ be g-variate first order autoregressive process,*

$$X_t = AX_{t-1} + c + \epsilon_t.$$

Assume that all eigenvalues of A are less than one in absolute value. $\{\epsilon_t\}$ is an i.i.d white noise process satisfying

$$E(\epsilon_t) = 0, \quad E(\epsilon_t \epsilon_s') = 0, \quad E(\epsilon_t \epsilon_t') = \Sigma, \text{ for all } t, s,$$

where Σ is a finite invertible matrix of constants. Then the following results hold:

1. *X_t has a vector MA(∞) representation,*

$$X_t = \mu + \sum_{j=0}^{\infty} B_j \epsilon_{t-j} \tag{B.1}$$

The sequence $\{B_j\}$ is absolutely summable and

$$\mu = (I - A)^{-1} c.$$

2. *The process $\{X_t\}$ is second order stationary. Its mean is given by*

$$E(X_t) = \mu,$$

the variance-covariance matrix satisfies

$$vec[Var(X_t)] = (I_{g^2} - A \otimes A)^{-1} vec(\Sigma) \tag{B.2}$$

3. Let $I = \{1, 2, \ldots, g\}$. Assume that for all $(i_1, i_2, i_3) \in I \times I \times I$ and all t, $E(\epsilon_{i_1,t}\epsilon_{i_2,t}\epsilon_{i_3,t})$ is a finite constant that does not depend on t, which we denote by

$$E(\epsilon_{i_1,t}\epsilon_{i_2,t}\epsilon_{i_3,t}) = g(i_1, i_2, i_3) < \infty.$$

Similarly, assume that

$$E(\epsilon_{i_1,t}\epsilon_{i_2,t}\epsilon_{i_3,t}\epsilon_{i_4,t}) = g(i_1, i_2, i_3, i_4) < \infty.$$

That is, we postulate time-invariant finite third and fourth moments of the white noise process $\{\epsilon_t\}$. Let $X_t^* = X_t - \mu$. Then for all $(i_1, i_2, i_3, i_4) \in I \times I \times I \times I$ and all t, $E(X_{i_1t}^* X_{i_2t,}^* X_{i_3t}^*)$ and $E(X_{i_1t}^* X_{i_2t}^* X_{i_3t}^* X_{i_4t}^*)$ are finite and time-invariant.

Proof. For 1. and 2. see [76]. For 3. we prove fourth order stationarity. The proof for third order stationarity is analogous. For notational convenience we assume $c = 0$, otherwise all X would have to carry asterisks.

Write the coefficient matrix of ϵ_{t-j} in the MA(∞) representation (B.1) as

$$B_j = \begin{pmatrix} B_{11,j} & \ldots & B_{1g,j} \\ \vdots & \ddots & \vdots \\ B_{g1,j} & \ldots & B_{gg,j} \end{pmatrix}.$$

Thus, the MA(∞) representation of an arbitrary X_{it} is

$$\begin{aligned} X_{it} &= B_{i1,0}\epsilon_{1,t} + B_{i2,0}\epsilon_{2,t} + \ldots + B_{ig,0}\epsilon_{g,t} \\ &\quad + B_{i1,1}\epsilon_{1,t-1} + B_{i2,1}\epsilon_{2,t-1} + \ldots + B_{ig,1}\epsilon_{g,t-1} \\ &\quad + B_{i1,2}\epsilon_{1,t-2} + B_{i2,2}\epsilon_{2,t-2} + \ldots + B_{ig,2}\epsilon_{g,t-2} \\ &\quad + \ldots \\ &= \sum_{j=0}^{\infty} B_{i1,j}\epsilon_{1,t-j} + \sum_{j=0}^{\infty} B_{i2,j}\epsilon_{2,t-j} + \ldots + \sum_{j=0}^{\infty} B_{ig,j}\epsilon_{g,t-j} \end{aligned}$$

Thus, for a fourth moment of X_t:

$$\begin{aligned} &E(X_{i_1t}X_{i_2t}X_{i_3t}X_{i_4t}) \\ &= E\left[\left(\sum_{j=0}^{\infty} B_{i_11,j}\epsilon_{1,t-j} + \sum_{j=0}^{\infty} B_{i_12,j}\epsilon_{2,t-j} + \ldots + \sum_{j=0}^{\infty} B_{i_1g,j}\epsilon_{g,t-j}\right)\right. \\ &\quad \times \left(\sum_{j=0}^{\infty} B_{i_21,j}\epsilon_{1,t-j} + \sum_{j=0}^{\infty} B_{i_22,j}\epsilon_{2,t-j} + \ldots + \sum_{j=0}^{\infty} B_{i_2g,j}\epsilon_{g,t-j}\right) \\ &\quad \times \left(\sum_{j=0}^{\infty} B_{i_31,j}\epsilon_{1,t-j} + \sum_{j=0}^{\infty} B_{i_32,j}\epsilon_{2,t-j} + \ldots + \sum_{j=0}^{\infty} B_{i_3g,j}\epsilon_{g,t-j}\right) \\ &\quad \times \left.\left(\sum_{j=0}^{\infty} B_{i_41,j}\epsilon_{1,t-j} + \sum_{j=0}^{\infty} B_{i_42,j}\epsilon_{2,t-j} + \ldots + \sum_{j=0}^{\infty} B_{i_4g,j}\epsilon_{g,t-j}\right)\right] \end{aligned}$$

Expanding yields a finite sum with g^4 summands. Each summand is a product of four infinite sums, one sum from each row. The following is a typical summand of this type, the index l_k denotes which of the g sums of the kth row is selected. We have:

$$
\begin{aligned}
E&\left[\left(\sum_{j=0}^{\infty} B_{i_1 l_1,j}\epsilon_{1,t-j}\right)\left(\sum_{j=0}^{\infty} B_{i_2 l_2,j}\epsilon_{1,t-j}\right)\right.\\
&\left.\left(\sum_{j=0}^{\infty} B_{i_3 l_3,j}\epsilon_{1,t-j}\right)\left(\sum_{j=0}^{\infty} B_{i_4 l_4,j}\epsilon_{1,t-j}\right)\right]\\
= E&\left[\sum_{j_1=0}^{\infty}\sum_{j_2=0}^{\infty}\sum_{j_3=0}^{\infty}\sum_{j_4=0}^{\infty} B_{i_1 l_1,j_1}\cdot B_{i_2 l_2,j_2}\cdot B_{i_3 l_3,j_3}\cdot B_{i_4 l_4,j_4}\right.\\
&\left.\cdot\epsilon_{l_1,t-j_1}\cdot\epsilon_{l_2,t-j_2}\cdot\epsilon_{l_3 t-j_3}\cdot\epsilon_{l_4,t-j_4}\right]\\
= &\sum_{j_1=0}^{\infty} B_{i_1 l_1,j_1}\sum_{j_2=0}^{\infty} B_{i_2 l_2,j_2}\sum_{j_3=0}^{\infty} B_{i_3 l_3,j_3}\sum_{j_4=0}^{\infty} B_{i_4 l_4,j_4}\\
&\cdot E(\epsilon_{l_1,t-j_1}\cdot\epsilon_{l_2,t-j_2}\cdot\epsilon_{l_3,t-j_3}\cdot\epsilon_{l_4,t-j_4})
\end{aligned}
$$

If there is at least one date among the four dates $t-j_1$, $t-j_2$, $t-j_3$, $t-j_4$, that is not equal to one of the other three, the expectation $E(\epsilon_{l_1,t-j_1}\cdot\epsilon_{l_2,t-j_2}\cdot\epsilon_{l_3,t-j_3}\cdot\epsilon_{l_4,t-j_4})$ is zero, otherwise it is equal to a finite constant that only depends on j_1, j_2, j_3 and j_4. Moreover, each of the sequences $B_{i_1 l_1,j_1}$, $B_{i_2 l_2,j_2}$, $B_{i_3 l_3,j_3}$, $B_{i_4 l_4,j_4}$ is absolutely summable, so the above expression is finite and depends on i_1, i_2, i_3 i_4, j_1, j_2, j_3, and j_4 only.

The only thing that remains to be shown is that we were allowed to interchange the summation and expectation operator for the second equality of the above expression. If $\{U_i\}$ is a sequence of random variables, then for $E(\sum_{i=0}^{\infty} U_i)=\sum_{i=1}^{\infty} E(U_i)$ to hold, it is sufficient that $\sum_{i=0}^{\infty} E(|U_i|)<\infty$.[1]

Here, we have to show that

$$
\begin{aligned}
&\sum_{j_1=0}^{\infty}\sum_{j_2=0}^{\infty}\sum_{j_3=0}^{\infty}\sum_{j_4=0}^{\infty} E|B_{i_1 l_1,j_1}\cdot B_{i_2 l_2,j_2}\cdot B_{i_3 l_3,j_3}\cdot B_{i_4 l_4,j_4}\\
&\cdot\epsilon_{l_1,t-j_1}\cdot\epsilon_{l_2,t-j_2}\cdot\epsilon_{l_3,t-j_3}\cdot\epsilon_{l_4,t-j_4}|\\
=&\sum_{j_1=0}^{\infty}\sum_{j_2=0}^{\infty}\sum_{j_3=0}^{\infty}\sum_{j_4=0}^{\infty}|B_{i_1 l_1,j_1}|\cdot|B_{i_2 l_2,j_2}|\cdot|B_{i_3 l_3,j_3}|\cdot|B_{i_4 l_4,j_4}|\\
&\cdot E|\epsilon_{l_1,t-j_1}\cdot\epsilon_{l_2,t-j_2}\cdot\epsilon_{l_3,t-j_3}\cdot\epsilon_{l_4,t-j_4}| \qquad \text{(B.3)}
\end{aligned}
$$

is finite. In order to prove that we use Hölder's inequality: for two random variables V and$W$:

[1] See, e.g. [53].

$$E|V \cdot W| \leq \sqrt{E(|V^2|) \cdot E(|V^2|)}.$$

Applying it to the above expectation yields

$$\begin{aligned}
& E|\epsilon_{l_1,t-j_1} \cdot \epsilon_{l_2,t-j_2} \cdot \epsilon_{l_3,t-j_3} \cdot \epsilon_{l_4,t-j_4}| \\
\leq & \sqrt{E(|\epsilon^2_{l_1,t-j_1} \cdot \epsilon^2_{l_2,t-j_2}|) \cdot E(|\epsilon^2_{l_3,t-j_3} \cdot \epsilon^2_{l_4,t-j_4}|)} \\
\leq & \sqrt{\sqrt{E(|\epsilon^4_{l_1,t-j_1}|) \cdot E(|\epsilon^4_{l_2,t-j_2}|)} \cdot \sqrt{E(|\epsilon^4_{l_3,t-j_3}|) \cdot E(|\epsilon^4_{l_4,t-j_4}|)}} \\
= & \left[E(\epsilon^4_{l_1,t-j_1}) \cdot E(\epsilon^4_{l_2,t-j_2}) \cdot E(\epsilon^4_{l_3,t-j_3}) \cdot E(\epsilon^4_{l_4,t-j_4})\right]^{\frac{1}{4}} \\
\leq & \infty
\end{aligned}$$

The first inequality uses Hölder's inequality with $V = \epsilon_{l_1,t-j_1} \cdot \epsilon_{l_2,t-j_2}$ and $W = \epsilon_{l_3,t-j_3} \cdot \epsilon_{l_4,t-j_4}$. The second inequality employs it twice, first with $V = \epsilon^2_{l_1,t-j_1}$ and $W = \epsilon^2_{l_2,t-j_2}$, second with $V = \epsilon^2_{l_3,t-j_3}$ and $W = \epsilon^2_{l_4,t-j_4}$.

Moreover, each of the four sequences of coefficient matrices in (B.3) is absolutely summable, so the whole expression (B.3) is finite. □

C

Derivations for the One-Factor Models in Discrete Time

C.1 Sharpe Ratios for the One-Factor Models

We derive the Sharpe Ratios for the discrete-time one-factor models from sections 3.2.1 and 3.2.3, respectively. In both cases we have:

$$\begin{aligned}
& r_{t+1}^n - r_{t+1}^f \\
&= \ln P_{t+1}^{n-1} - \ln P_t^n + \ln P_t^1 \\
&= -A_{n-1} - B_{n-1}X_{t+1} + A_n + B_n X_t - X_t \\
&= A_n - A_{n-1} - B_{n-1}\theta(1-\kappa) \\
&\quad +(B_{n-1}\kappa - 1 + B_n)X_t + B_{n-1}u_{t+1}.
\end{aligned}$$

It is easy to see that $B_{n-1}\kappa - 1 + B_n = 0$. For the remainder of the expression we have to distinguish between the two models. For the Vasiček model:

$$\begin{aligned}
& A_n - A_{n-1} - B_{n-1}\theta(1-\kappa) \\
&= \delta + B_{n-1}\theta(1-\kappa) - \frac{1}{2}(\lambda + B_{n-1})^2\sigma^2 - B_{n-1}\theta(1-\kappa) \\
&= \frac{1}{2}\lambda^2\sigma^2 - \frac{1}{2}\lambda^2\sigma^2 - B_{n-1}\lambda\sigma^2 - \frac{1}{2}B_{n-1}^2\sigma^2 \\
&= -B_{n-1}\lambda\sigma^2 - \frac{1}{2}B_{n-1}^2\sigma^2.
\end{aligned}$$

For the mixture model:

$$\begin{aligned}
& A_n - A_{n-1} - B_{n-1}\theta(1-\kappa) \\
&= \delta + B_{n-1}\theta(1-\kappa) \\
&\quad - \ln\left(\omega e^{\frac{1}{2}(\lambda+B_{n-1})^2\sigma_1^2} + (1-\omega)e^{\frac{1}{2}(\lambda+B_{n-1})^2\sigma_2^2}\right) - B_{n-1}\theta(1-\kappa) \\
&= \ln\left(\omega e^{\frac{1}{2}\lambda^2\sigma_1^2} + (1-\omega)e^{\frac{1}{2}\lambda^2\sigma_2^2}\right) \\
&\quad - \ln\left(\omega e^{\frac{1}{2}(\lambda+B_{n-1})^2\sigma_1^2} + (1-\omega)e^{\frac{1}{2}(\lambda+B_{n-1})^2\sigma_2^2}\right)
\end{aligned}$$

For both models $E_t(B_{n-1}u_{t+1}) = 0$ and $Var_t(B_{n-1}u_{t+1}) = B_{n-1}^2\sigma^2$, where for the mixture model $\sigma^2 := \omega\sigma_1^2 + (1-\omega)\sigma_2^2$.

Thus, we obtain for the Vasiček model

$$SR_t^n = \frac{E_t(r_{t+1}^n - r_{t+1}^f)}{\sqrt{Var_t(r_{t+1}^n - r_{t+1}^f)}} \tag{C.1}$$

$$= \frac{-B_{n-1}\lambda\sigma^2 - \frac{1}{2}B_{n-1}^2\sigma^2}{B_{n-1}\sigma}, \tag{C.2}$$

and for the mixture model

$$SR_t^n = \frac{E_t(r_{t+1}^n - r_{t+1}^f)}{\sqrt{Var_t(r_{t+1}^n - r_{t+1}^f)}} \tag{C.3}$$

$$= \frac{1}{B_{n-1}\sigma}\Big[\ln\left(\omega e^{\frac{1}{2}\lambda^2\sigma_1^2} + (1-\omega)e^{\frac{1}{2}\lambda^2\sigma_2^2}\right) \tag{C.4}$$

$$- \ln\left(\omega e^{\frac{1}{2}(\lambda+B_{n-1})^2\sigma_1^2} + (1-\omega)e^{\frac{1}{2}(\lambda+B_{n-1})^2\sigma_2^2}\right)\Big]$$

We now show that SR_t^n for the mixture model is monotonic decreasing in λ. We take the derivative with respect to λ and show that the resulting expression is negative. Since the result holds for arbitrary n we choose $n+1$ for notational convenience.

$$\frac{\partial SR_t^{n+1}}{\partial\lambda} = \frac{1}{B_n\sigma}\left[\frac{\omega e^{\frac{1}{2}\lambda^2\sigma_1^2}\sigma_1^2\lambda + (1-\omega)e^{\frac{1}{2}\lambda^2\sigma_2^2}\sigma_2^2\lambda}{\omega e^{\frac{1}{2}\lambda^2\sigma_1^2} + (1-\omega)e^{\frac{1}{2}\lambda^2\sigma_2^2}}\right.$$

$$\left. - \frac{\omega e^{\frac{1}{2}(\lambda+B_n)^2\sigma_1^2}\sigma_1^2(\lambda+B_n) + (1-\omega)e^{\frac{1}{2}(\lambda+B_n)^2\sigma_2^2}\sigma_2^2(\lambda+B_n)}{\omega e^{\frac{1}{2}(\lambda+B_n)^2\sigma_1^2} + (1-\omega)e^{\frac{1}{2}(\lambda+B_n)^2\sigma_2^2}}\right].$$

The product of the denominators of the two fractions in the difference is clearly positive. Writing the difference as one fraction we obtain for the enumerator after expanding terms:

$$\begin{aligned}
&\omega e^{\frac{1}{2}\lambda^2\sigma_1^2}\sigma_1^2\lambda \cdot \omega e^{\frac{1}{2}(\lambda+B_n)^2\sigma_1^2} \\
&+\omega e^{\frac{1}{2}\lambda^2\sigma_1^2}\sigma_1^2\lambda \cdot (1-\omega)e^{\frac{1}{2}(\lambda+B_n)^2\sigma_2^2} \\
&+(1-\omega)e^{\frac{1}{2}\lambda^2\sigma_2^2}\sigma_2^2\lambda \cdot \omega e^{\frac{1}{2}(\lambda+B_n)^2\sigma_1^2} \\
&+(1-\omega)e^{\frac{1}{2}\lambda^2\sigma_2^2}\sigma_2^2\lambda \cdot (1-\omega)e^{\frac{1}{2}(\lambda+B_n)^2\sigma_2^2} \\
&-\omega e^{\frac{1}{2}(\lambda+B_n)^2\sigma_1^2}\sigma_1^2(\lambda+B_n) \cdot \omega e^{\frac{1}{2}\lambda^2\sigma_1^2} \\
&-\omega e^{\frac{1}{2}(\lambda+B_n)^2\sigma_1^2}\sigma_1^2(\lambda+B_n) \cdot (1-\omega)e^{\frac{1}{2}\lambda^2\sigma_2^2} \\
&-(1-\omega)e^{\frac{1}{2}(\lambda+B_n)^2\sigma_2^2}\sigma_2^2(\lambda+B_n) \cdot \omega e^{\frac{1}{2}\lambda^2\sigma_1^2} \\
&-(1-\omega)e^{\frac{1}{2}(\lambda+B_n)^2\sigma_2^2}\sigma_2^2(\lambda+B_n) \cdot (1-\omega)e^{\frac{1}{2}\lambda^2\sigma_2^2}
\end{aligned}$$

$$\begin{aligned}
&= \omega^2 e^{\frac{1}{2}(\lambda+B_n)^2\sigma_1^2+\frac{1}{2}\lambda^2\sigma_1^2} \cdot [\sigma_1^2\lambda - \sigma_1^2(\lambda+B_n)] \\
&+\omega(1-\omega)e^{\frac{1}{2}(\lambda+B_n)^2\sigma_1^2+\frac{1}{2}\lambda^2\sigma_2^2} \cdot [\sigma_1^2\lambda - \sigma_2^2(\lambda+B_n)] \\
&+\omega(1-\omega)e^{\frac{1}{2}(\lambda+B_n)^2\sigma_1^2+\frac{1}{2}\lambda^2\sigma_2^2} \cdot [\sigma_2^2\lambda - \sigma_1^2(\lambda+B_n)] \\
&+(1-\omega)^2 e^{\frac{1}{2}(\lambda+B_n)^2\sigma_2^2+\frac{1}{2}\lambda^2\sigma_2^2} \cdot [\sigma_2^2\lambda - \sigma_2^2(\lambda+B_n)]
\end{aligned}$$

Obviously, the first and the fourth addend of the last expression are negative. The sum of the second and third addend simplifies to

$$\omega(1-\omega)e^{\frac{1}{2}\lambda+B_n)^2\sigma_1^2+\frac{1}{2}\lambda^2\sigma_2^2} \cdot [-(\sigma_1^2+\sigma_2^2)B_n]$$

which is also negative. Therefore, the Sharpe ratio for the mixture model is decreasing in λ.

C.2 The Kurtosis Increases in the Variance Ratio

We show that the kurtosis of u_t rises in the ratio of variances. Let $c > 1$ and $\sigma_1^2 = c\sigma_2^2$. We have

$$\begin{aligned}
&kurt(u_t) \\
&= \frac{\omega 3\sigma_1^4 + (1-\omega)3\sigma_2^4}{[\omega\sigma_1^2 + (1-\omega)\sigma_2^2]^2} - 3 \\
&= \frac{3[\omega c^2\sigma_2^4 + (1-\omega)\sigma_2^4]}{[\omega c\sigma_2^2 + (1-\omega)\sigma_2^2]^2} - 3 \\
&= 3\frac{\omega c^2 + 1 - \omega}{(\omega c + 1 - \omega)^2} - 3 \\
&=: H(c)
\end{aligned}$$

The derivative with respect to the variance ratio is

$$\begin{aligned}
&\frac{\partial H(c)}{\partial c}\\
&= 3\frac{2\omega c(\omega c+1-\omega)^2-(\omega c^2+1-\omega)\cdot 2(\omega c+1-\omega)\omega}{(\omega c+1-\omega)^4}\\
&= 3\frac{2\omega c(\omega c+1-\omega)[c(\omega c+1-\omega)-\omega c^2+1-\omega]}{(\omega c+1-\omega)^4}\\
&= 3\frac{2\omega c(\omega c+1-\omega)[c(1-\omega)+(1-\omega)]}{(\omega c+1-\omega)^4},
\end{aligned}$$

which is clearly positive.

C.3 Derivation of Formula (3.53)

We show that the third moment of u_t can be written as (3.53):

$$\begin{aligned}
&E(u_t^3)\\
&= \omega_1(3\mu_1\sigma_1^2+\mu_1^3)+\omega_2(3\mu_2\sigma_2^2+\mu_2^3)\\
&= \omega_1(3\mu_1\sigma_1^2+\mu_1^3)+\omega_2\left(3\left(-\frac{\omega_1}{\omega_2}\mu_1\right)\sigma_2^2-\left(\frac{\omega_1}{\omega_2}\mu_1\right)^3\right)\\
&= \omega_1 3\mu_1\sigma_1^2-\omega_1 3\mu_1\sigma_2^2+\omega_1\mu_1^3-\omega_1^3\omega_2^2\mu_1^3\\
&= \omega_1\mu_1\left(3(\sigma_1^2-\sigma_2^2)+\frac{\omega_2-\omega_1}{\omega_2^2}\mu_1^2\right).
\end{aligned}$$

Here the second equality uses $\mu_2 = -\omega_1/\omega_2\mu_1$, which follows from $\omega_1\mu_1 + \omega_2\mu_2 = 0$. The last equality uses

$$1-\frac{\omega_1^2}{\omega_2^2}=\frac{\omega_2^2-\omega_1^2}{\omega_2^2}=\frac{(\omega_1+\omega_2)(\omega_2-\omega_1)}{\omega_2^2}=\frac{\omega_2-\omega_1}{\omega_2^2}.$$

C.4 Moments of Factors

The following derivations make use of stationarity up to fourth order of the state process. Furthermore, the independence between powers of X_t^* and u_t is utilized.

Second moment:

$$\begin{aligned}
E(X_t^{*2}) &= E[(\kappa X_{t-1}^*+u_t)^2]\\
&= \kappa^2E(X_{t-1}^{*2})+2\kappa E(X_{t-1}^*u_t)+E(u_t^2)\\
&= \kappa^2E(X_t^{*2})+2\kappa E(X_{t-1}^*)E(u_t)+\sigma^2\\
\Rightarrow E(X_t^{*2}) &= \frac{\sigma^2}{1-\kappa^2}
\end{aligned}$$

Third moment:

$$\begin{aligned}
E(X_t^{*3}) &= E[(\kappa X_{t-1}^* + u_t)^3] \\
&= \kappa^3 E(X_{t-1}^{*3}) + 3\kappa^2 E(X_{t-1}^{*2} u_t) + 3\kappa E(X_{t-1}^* u_t^2) + E(u_t^3) \\
&= \kappa^3 E(X_t^{*3}) + E(u_t^3) \\
\Rightarrow E(X_t^{*3}) &= \frac{E(u_t^3)}{1-\kappa^3}
\end{aligned}$$

Fourth moment:

$$\begin{aligned}
&E(X_t^{*4}) \\
&= E[(\kappa X_{t-1}^* + u_t)^4] \\
&= \kappa^4 E(X_{t-1}^{*4}) + 4\kappa^3 E(X_{t-1}^{*3} u_t) + 6\kappa^2 E(X_{t-1}^{*2} u_t^2) + 4\kappa E(X_{t-1}^* u_t^3) + E(u_t^4) \\
&= \kappa^4 E(X_t^{*4}) + 6\kappa^2 E(X_t^{*2}) E(u_t^2) + E(u_t^4)
\end{aligned}$$

$$\Rightarrow E(X_t^{*4}) = \frac{6\kappa^2\sigma^4}{(1-\kappa^2)(1-\kappa^4)} + \frac{E(u_t^4)}{1-\kappa^4} = \frac{6\kappa^2\sigma^4 + (1-\kappa^2)E(u_t^4)}{(1-\kappa^2)(1-\kappa^4)}$$

C.5 Skewness and Kurtosis of Yields

We derive the formulas for the skewness and kurtosis of y_t^n:

$$\begin{aligned}
skew(y_t^n) &= \frac{\left(\frac{B_n}{n}\right)^3 E[(X_t^*)^3]}{\left(\frac{B_n}{n}\right)^{2\frac{3}{2}} (E[(X_t^*)^2])^{\frac{3}{2}}} \\
&= \frac{E(u_t^3)}{1-\kappa^3} \cdot \frac{(1-\kappa^2)^{\frac{3}{2}}}{\sigma^3} \\
&= \frac{(1-\kappa^2)^{\frac{3}{2}}}{1-\kappa^3} skew(u_t),
\end{aligned}$$

$$
\begin{aligned}
kurt(y_t^n) &= \frac{\left(\frac{B_n}{n}\right)^4 E[(X_t^*)^4]}{\left(\frac{B_n}{n}\right)^4 (E[(X_t^*)^2])^2} - 3 \\
&= \frac{[6\kappa^2\sigma^4 + (1-\kappa^2)E(u_t^4)](1-\kappa^2)^2}{(1-\kappa^2)(1-\kappa^4)\sigma^4} - 3 \\
&= \frac{6\kappa^2\sigma^4 + (1-\kappa^2)E(u_t^4) - 3(1+\kappa^2)\sigma^4}{(1+\kappa^2)\sigma^4} \\
&= \frac{-3(1-\kappa^2)\sigma^4 + (1-\kappa^2)E(u_t^4)}{(1+\kappa^2)\sigma^4} \\
&= \frac{(1-\kappa)^2(E(u_t^4) - 3\sigma^4)}{(1+\kappa^2)\sigma^4} \\
&= \frac{(1-\kappa)^2\sigma^4 kurt(u_t)}{(1+\kappa^2)\sigma^4} \\
&= \frac{(1-\kappa)^2}{1+\kappa^2} kurt(u_t)
\end{aligned}
$$

The kurtosis of y_t^n is strictly decreasing in κ. We have

$$
\begin{aligned}
&\frac{\partial}{\partial\kappa}\frac{(1-\kappa)^2}{1+\kappa^2} \\
&= \frac{-2(1-\kappa)(1+\kappa^2) - 2\kappa(1-\kappa)^2}{(1+\kappa^2)^2} \\
&= \frac{-2(1-\kappa)[1+\kappa^2+\kappa(1-\kappa)]}{(1+\kappa)^2} \\
&= \frac{-2(1-\kappa^2)}{(1+\kappa^2)^2},
\end{aligned}
$$

which is clearly negative.

C.6 Moments of Differenced Factors

We compute moments of the factor process in first differences. The results are given in terms of moments of the state process in levels and in moments of u_t, which is convenient for the computation of the variance, the skewness, and the kurtosis of the Δy_t^n.

In the following we will use that

$$
\begin{aligned}
\Delta X_t &= X_t - X_{t-1} \\
&= \theta + \kappa(X_{t-1} - \theta) + u_t - X_{t-1} \\
&= -(1-\kappa)X_{t-1}^* + u_t.
\end{aligned}
$$

Second moment:

$$\begin{aligned}
E[(\Delta X_t)^2] &= E[(-(1-\kappa)X^*_{t-1} + u_t)^2] \\
&= (1-\kappa)^2 E(X_t^{*2}) + \sigma^2 \\
&= \frac{(1-\kappa)^2\sigma^2}{1-\kappa^2} + \sigma^2 \\
&= \left(\frac{(1-\kappa)}{(1+\kappa)} + 1\right)\sigma^2 \\
&= \frac{2}{1+\kappa}\sigma^2
\end{aligned}$$

Third moment:

$$\begin{aligned}
E[(\Delta X_t)^3] &= E[(-(1-\kappa)X^*_{t-1} + u_t)^3] \\
&= -(1-\kappa)^3 E(X_t^{*3}) + E(u_t^3) \\
&= \frac{-(1-\kappa)^3}{1-\kappa^3} E(u_t^3) + E(u_t^3) \\
&= \frac{(\kappa-1)^3 + (1-\kappa^3)}{1-\kappa^3} E(u_t^3) \\
&= \frac{3\kappa(1-\kappa)}{1-\kappa^3} E(u_t^3)
\end{aligned}$$

Fourth moment:

$$\begin{aligned}
E[(\Delta X_t)^4] &= E[(-(1-\kappa)X^*_{t-1} + u_t)^4] \\
&= (1-\kappa)^4 E(X_t^{*4}) + 6(1-\kappa)^2 E(X_t^{*2})\sigma^2 + E(u_t^4).
\end{aligned}$$

C.7 Moments of Differenced Yields

Using the moments of the differenced factor process, we now compute the variance, the coefficient of skewness and the coefficient of kurtosis for yield changes:

Variance:

$$\begin{aligned}
E[(\Delta y_t^n)^2] &= \left(\frac{B_n}{n}\right)^2 E[(\Delta X_t)^2] \\
&= \left(\frac{B_n}{n}\right)^2 \frac{2}{1+\kappa}\sigma^2.
\end{aligned}$$

Skewness:

$$\begin{aligned}
skew(\Delta y_t^n) &= \frac{\left(\frac{B_n}{n}\right)^3 E[(\Delta X_t^{n*})^3]}{\left(\frac{B_n}{n}\right)^{2\frac{3}{2}} (E[(\Delta X_t^{n*})^2])^{\frac{3}{2}}} \\
&= \frac{3\kappa(1-\kappa)}{1-\kappa^3} E(u_t^3) \cdot \frac{(1+\kappa)^{\frac{3}{2}}}{2^{\frac{3}{2}}\sigma^3} \\
&= \frac{3\kappa(1-\kappa)(1+\kappa)^{\frac{3}{2}}}{(1-\kappa^3)2^{\frac{3}{2}}} skew(u_t)
\end{aligned}$$

Kurtosis:

$$kurt(\Delta y_t^n) = \frac{\left(\frac{B_n}{n}\right)^4 E[(\Delta X_t^n)^4]}{\left(\frac{B_n}{n}\right)^4 (E[(\Delta X_t^n)^2])^2} - 3$$
$$= \frac{(1-\kappa)^4 E(X_t^{*4}) + 6(1-\kappa)^2 E(X_t^{*2})\sigma^2 + E(u_t^4)}{\frac{4\sigma^4}{(1+\kappa)^2}} - 3$$

We consider the three summands of the sum separately. For the first summand:

$$\frac{(1-\kappa)^4 E(X_t^{*4})(1+\kappa)^2}{4\sigma^4}$$
$$= \frac{(1-\kappa)^4 E(X_t^{*4})(1+\kappa)^2}{4[E(X_t^{*2})]^2(1-\kappa^2)^2}$$
$$= \frac{(1-\kappa)^2}{4}\left[\frac{(1-\kappa)^2}{1+\kappa^2} kurt(u_t) + 3\right]$$

For the second summand:

$$\frac{6(1-\kappa)^2 E(X_t^{*2})\sigma^2(1+\kappa)^2}{4\sigma^4}$$
$$= \frac{6(1-\kappa)^2 E(X_t^{*2})(1+\kappa)^2}{4(1-\kappa^2)E(X_t^{*2})}$$
$$= \frac{3}{2}\frac{[(1-\kappa)(1+\kappa)]^2}{1-\kappa^2}$$
$$= \frac{3}{2}(1-\kappa^2)$$

And for the third summand:

$$\frac{E(u_t^4)(1+\kappa)^2}{4\sigma^4}$$
$$= [kurt(u_t) + 3]\frac{(1+\kappa)^2}{4}$$

Collecting the coefficients of $kurt(u_t)$ and the remaining terms:

$$kurt(\Delta y_t^n) = \frac{(1-\kappa)^4 + (1+\kappa^2)(1+\kappa)^2}{4(1+\kappa^2)} kurt(u_t)$$
$$+\frac{3}{4}(1-\kappa)^2 + \frac{3}{2}(1-\kappa^2) + \frac{3}{4}(1+\kappa)^2 - 3$$
$$= \frac{(1-\kappa)^4 + (1+\kappa^2)(1+\kappa)^2}{4(1+\kappa^2)} kurt(u_t)$$
$$+ \frac{3[1 - 2\kappa + \kappa^2 + 2 - 2\kappa^2 + \kappa^2 + 2\kappa + 1] - 12}{4}$$
$$= \frac{(1-\kappa)^4 + (1+\kappa^2)(1+\kappa)^2}{4(1+\kappa^2)} kurt(u_t) + 0$$

D

A Note on Scaling

In this section, we analyze if the Gaussian one-factor Vasiček model is invariant with respect to a rescaling of the vector of yields. Let $\{n_1, \dots n_k\}$ be an arbitrary collection of maturities. The joint evolution of $(y_t^{n_1}, \dots, y_t^{n_k})'$ is fully specified by the functional relation between y_t^n and X_t, (3.24),

$$y_t^n = \frac{A_n}{n} + \frac{B_n}{n} X_t,$$

and the evolution of the state variable X_t as an AR(1) process, (3.20),[1]

$$X_t = \theta + \kappa(X_{t-1} - \theta) + u_t.$$

Given the functional forms and the specification of the distribution of u_t in (3.22) as a mean zero normal distribution, the model is fully described by the choice of the parameters $\psi = (\theta, \kappa, \sigma^2, \delta, \lambda)'$.[2] In order to focus on the dependence of the parameters, we write the model for an arbitrary n as

$$y_t^n = f_A(\psi, n) + f_B(\psi, n) \cdot X_t, \tag{D.1}$$
$$X_t = f_X(\psi, X_{t-1}) + u_t, \quad u_t \sim N(0, f_\sigma(\psi)). \tag{D.2}$$

For instance, $f_A(\psi, n)$ represents the intercept term A_n/n in (3.24) explicitly as a function of the model parameters and time to maturity. We now claim that if the joint evolution of y_t^n and X_t is given by (D.1) - (D.2), then the same model, i.e. with the same functional forms but possibly different parameters, is not valid for the evolution of the scaled yield $\tilde{y}_t^n := c \cdot y_t^n$.

This implies that one has to be careful with respect to the scaling of yields in empirical applications. The theoretical model holds for monthly yields, i.e. an annualized two-year yield of 6% would be measured as $y_t^{24} = 6/1200 =$

[1] One may also specify an initial condition for the factor process $\{X_t\}$, but this does not change the following argument.

[2] Note that we treat δ here as a free parameter. That is, we do not impose the condition that X_t equals y_t^1.

0.005. Estimating the model parameters with scaled data (i.e. having the latter yield as 6 instead of 0.005 in the data set) leads to parameter estimates that cannot be interpreted as those of the original model for monthly yields. As a solution to this problem one either has to modify the functional form of A_n (as shown below) or to simply multiply through (D.1) by the scaling factor (as done in the empirical application in chapter 9).

Next we formally state the scaling problem and then propose a solution.

Proposition D.1. *For $c > 0$ define $\tilde{y}_t^n := c \cdot y_t^n$ for all n and t. Then the model (D.1) - (D.2) is not valid for $\tilde{y}_t^n$ in the sense that there does not exist a parameter vector $\tilde{\psi}$, a factor process $\{\tilde{X}_t\}$ and a factor innovation $\tilde{u}_t$ such that*

$$\tilde{y}_t^n = f_A(\tilde{\psi}, n) + f_B(\tilde{\psi}, n) \cdot \tilde{X}_t, \tag{D.3}$$

$$\tilde{X}_t = f_X(\tilde{\psi}, \tilde{X}_{t-1}) + \tilde{u}_t, \quad \tilde{u}_t \sim N(0, f_\sigma(\tilde{\psi})), \tag{D.4}$$

holds almost surely for all n and t.

Proof. We prove this by contradiction, and thus assume that such $\tilde{\psi}$, $\{\tilde{X}_t\}$ and $\tilde{u}_t$ do exist. Due to this assumption we can write for $n = 1$

$$\tilde{y}_t^1 = \tilde{\delta} - \frac{1}{2}\tilde{\lambda}^2\tilde{\sigma}^2 + \tilde{X}_t. \tag{D.5}$$

Since $\tilde{y}_t^n = c \cdot y_t^n$, we also have

$$\tilde{y}_t^1 = c\delta - \frac{1}{2}c\lambda^2\sigma^2 + cX_t,$$

which is just (D.1) multiplied through by c.

For the right hand sides to be identical for all X_t and $\tilde{X}_t$ we must have

$$\tilde{X}_t = c \cdot X_t. \tag{D.6}$$

From (D.2) in the original model it follows that

$$cX_t = c\theta + c\kappa(X_{t-1} - \theta) + cu_t.$$

Thus, the transformed factor satisfies

$$\tilde{X}_t = c\theta + \kappa(\tilde{X}_{t-1} - c\theta) + cu_t, \tag{D.7}$$

and one obtains for the parameters of the transformed model

$$\tilde{\theta} = c\theta, \quad \tilde{\kappa} = \kappa, \quad \tilde{u}_t = cu_t, \quad \tilde{\sigma}^2 = c^2\sigma^2. \tag{D.8}$$

As an immediate consequence one obtains that

$$f_B(\tilde{\psi}, n) = \frac{1 - \tilde{\kappa}^n}{1 - \tilde{\kappa}} = \frac{1 - \kappa^n}{1 - \kappa} = f_B(\psi, n) \tag{D.9}$$

that is, $\tilde{B}_n = B_n$.

For an arbitrary n we take expectations in (D.1) and (D.3). Since $E(\tilde{y}_t^n) = cE(y_t^n)$, one obtains the condition that

$$f_A(\tilde{\psi}, n) + f_B(\tilde{\psi}, n)\tilde{\theta} = c f_A(\psi, n) + f_B(\psi, n) c\theta. \tag{D.10}$$

Using (D.9), this leads to the requirement

$$f_A(\tilde{\psi}, n) = c f_A(\psi, n). \tag{D.11}$$

For $n = 1$ this becomes:

$$\tilde{\delta} - c^2 \frac{1}{2} \tilde{\lambda}^2 \sigma^2 = c\delta - c\frac{1}{2}\lambda^2\sigma^2 \tag{D.12}$$

Subtracting equation (D.11) for an arbitrary $n > 1$ from the same equation for $n+1$, and making use of (3.26) as well as (D.8), yields

$$\tilde{\delta} + B_n c\theta(1-\kappa) - \frac{1}{2}(\tilde{\lambda} + B_n)^2 c^2 \sigma^2 = c\delta + cB_n\theta(1-\kappa) - c\frac{1}{2}(\lambda + B_n)^2\sigma^2,$$

which can be rewritten as

$$\tilde{\delta} - c^2\frac{1}{2}\tilde{\lambda}^2\sigma^2 - c^2\frac{1}{2}(2B_n\tilde{\lambda} + B_n^2)\sigma^2 = c\delta - c\frac{1}{2}\lambda^2\sigma^2 - c\frac{1}{2}(2B_n\lambda + B_n^2)\sigma^2.$$

Using (D.12) from above, this becomes

$$c^2(2B_n\tilde{\lambda} + B_n^2)\sigma^2 = c(2B_n\lambda + B_n^2)\sigma^2.$$

Solving for $\tilde{\lambda}$ gives

$$\tilde{\lambda} = \frac{\lambda}{c} + \frac{1-c}{2c} B_n. \tag{D.13}$$

This depends on n and thus violates the assumption that $\tilde{\lambda}$ is a fixed parameter. □

We have shown that if the joint evolution of a vector of yields, $y_t = (y_t^{n_1}, \ldots, y_t^{n_k})'$, and a factor X_t satisfies (D.1) - (D.2), then there does not exist another set of parameters $\tilde{\psi}$ and a process $\{\tilde{X}_t\}$ such that the joint evolution of cy_t and $\tilde{X}_t$ satisfies (D.1) - (D.2) under this new parameterization. As evident from the proof, this property is a consequence of the way in which the parameters enter the intercept $A_n/n = f_A(\psi, n)$.

However, it is possible to provide another model that does describe the evolution of the vector process of scaled yields. It turns out that, compared to the original model, only the function f_A has to be modified slightly.

Proposition D.2 (The evolution of scaled yields). *For an arbitrary maturity n, let the joint evolution of the yield y_t^n and the factor X_t be described by (3.20), (3.22) and (3.24). Denote by $\tilde{y}_t^n := cy_t^n$ the scaled yield. Then there* *is a new factor process* $\{\tilde{X}_t\}$ *with* $\tilde{X}_t = cX_t$*, such that the joint evolution of* $\tilde{y}_t^n$ *and* $\tilde{X}_t$ *satisfies*

$$\tilde{y}_t^n = \frac{\tilde{A}_n^c}{n} + \frac{\tilde{B}_n}{n}\tilde{X}_t, \quad n = n_1, \ldots, n_k, \tag{D.14}$$

$$\tilde{X}_t = \tilde{\theta}_t + \tilde{\kappa}(\tilde{X}_{t-1} - \tilde{\theta}) + \tilde{u}_t, \quad \tilde{u}_t \sim i.i.d.\, N(0, \tilde{\sigma}^2), \tag{D.15}$$

where the new parameters in terms of the old ones are given as

$$\tilde{\theta} = c\theta, \quad \tilde{\kappa} = \kappa, \quad \tilde{u}_t = cu_t, \quad \tilde{\sigma}^2 = c^2\sigma^2, \quad \tilde{\delta} = c\delta, \quad \tilde{\lambda} = \lambda$$

and the coefficient functions satisfy

$$\tilde{B}_n = \frac{1 - \tilde{\kappa}^n}{1 - \tilde{\kappa}} \tag{D.16}$$

$$\tilde{A}_n^c = \sum_{i=0}^{n-1} \tilde{\delta} + \tilde{B}_i(\tilde{\theta}(1 - \tilde{\kappa})) - \frac{1}{2}(\tilde{\lambda} + \tilde{B}_i)^2 \frac{1}{c}\tilde{\sigma}^2. \tag{D.17}$$

Proof. That the transformed factor $\tilde{X}_t$ satisfies (D.15) is already clear from the proof of the last proposition. It remains to be shown that the scaled yield satisfies (D.14). We have

$$\begin{aligned}
&\tilde{y}_t^n \\
&= \frac{1}{n}\left[\sum_{i=0}^{n-1} \tilde{\delta} + \tilde{B}_i(\tilde{\theta}(1 - \tilde{\kappa})) - \frac{1}{2}(\tilde{\lambda} + \tilde{B}_i)^2 \frac{1}{c}\tilde{\sigma}^2\right] + \frac{1}{n}\frac{1 - \tilde{\kappa}^n}{1 - \tilde{\kappa}}\tilde{X}_t \\
&= \frac{1}{n}\left[\sum_{i=0}^{n-1} c\delta + B_i(c\theta(1 - \kappa)) - \frac{1}{2}(\lambda + B_i)^2 c\sigma^2\right] + \frac{1}{n}\frac{1 - \kappa^n}{1 - \kappa}cX_t \\
&= c\frac{A_n}{n} + c\frac{B_n}{n}X_t \\
&= cy_t^n.
\end{aligned}$$

□

The factor is scaled by the same coefficient c as yields themselves. The factor loading B_n/n does not change for the new parameterization. The only difference between $\tilde{A}_n/n = f_A(\tilde{\psi}, n)$ and A_n^c/n is the inclusion of the factor $1/c$ in (D.17). Note that this cannot be made disappear by absorbing it into one of the new parameters. If this was the case, we would in fact have the old model structure for the scaled yields, which, however, has been proved above to be unfeasible.

E

Derivations for the Multifactor Models in Discrete Time

E.1 Properties of Factor Innovations

Consider the distribution of the vector u_t of factor innovations:

$$u_t \sim \sum_{b=1}^{B} \omega_b N(\mu_b, V_b).$$

In the following we will repeatedly use moments of powers of u_t. For the following, let $M(u_{it}^m; b)$ denote the mth central moment around zero of the simple normal $N(\mu_{ib}, v_{ib}^2)$. That is, e.g.,

$$M(u_{it}^m; b) = \int u_{it}^m \, \phi(u_{it};\, \mu_{ib}, v_{ib}^2) du_{it},$$

As a consequence of the diagonality of the V_b we have the following result:

$$E(u_{it}^m) = \sum_{b=1}^{B} \omega_b \cdot M(u_{it}^m; b), \tag{E.1}$$

$$E(u_{it}^m \cdot u_{jt}^n) = \sum_{b=1}^{B} \omega_b \cdot M(u_{it}^m; b) \cdot M(u_{jt}^n; b), \tag{E.2}$$

$$E(u_{it}^m \cdot u_{jt}^n \cdot u_{kt}^p) = \sum_{b=1}^{B} \omega_b \cdot M(u_{it}^m; b) \cdot M(u_{jt}^n; b) \cdot M(u_{kt}^p; b), \tag{E.3}$$

$$E(u_{it}^m \cdot u_{jt}^n \cdot u_{kt}^p \cdot u_{lt}^q) = \sum_{b=1}^{B} \omega_b \cdot M(u_{it}^m; b) \cdot M(u_{jt}^n; b) \cdot M(u_{kt}^p; b) \cdot M(u_{lt}^q; b), \tag{E.4}$$

where for each case the indices i, j, k, l, are pairwise unequal. We prove the third relationship, the other results hold by the same argument. From section

3.2.2 we know, that given the density of $(u_{1t}, \dots u_{dt})'$, the density of any subvector, say $(u_{it}, u_{jt}, u_{kt})'$ is also a mixture of normals. We have

$$\begin{aligned} p(u_{it}, u_{jt}, u_{kt}) &= \sum_{b=1}^{B} \omega_b \phi \left(\begin{pmatrix} u_{it} \\ u_{jt} \\ u_{kt} \end{pmatrix}; \begin{pmatrix} \mu_{ib} \\ \mu_{jb} \\ \mu_{kb} \end{pmatrix}, \begin{pmatrix} v_{ib}^2 & 0 & 0 \\ 0 & v_{jb}^2 & 0 \\ 0 & 0 & v_{kb}^2 \end{pmatrix} \right) \\ &= \sum_{b=1}^{B} \omega_b \cdot \phi(u_{it}; \mu_{ib}, v_{ib}^2) \cdot \phi(u_{jt}; \mu_{jb}, v_{jb}^2) \cdot \phi(u_{kt}; \mu_{kb}, v_{kb}^2). \end{aligned}$$

Thus,

$$\begin{aligned} & E(u_{it}^m \cdot u_{jt}^n \cdot u_{kt}^p) \\ &= \int\int\int u_{it}^m \cdot u_{jt}^n \cdot u_{kt}^p \cdot p(u_{it}, u_{jt}, u_{kt}) du_{it} du_{jt} du_{kt} \\ &= \sum_{b=1}^{B} \omega_b \cdot \int u_{it}^m \phi(u_{it}; \mu_{ib}, v_{ib}^2) du_{it} \cdot \int u_{jt}^n \phi(u_{jt}; \mu_{jb}, v_{jb}^2) du_{jt} \\ & \quad \cdot \int u_{kt}^p \phi(u_{kt}; \mu_{kb}, v_{kb}^2) du_{kt} \\ &= \sum_{b=1}^{B} \omega_b \cdot M(u_{it}^m; b) \cdot M(u_{jt}^n; b) \cdot M(u_{kt}^p; b). \end{aligned}$$

From the fact that the u_t are independent over time, it follows that functions of elements of X_t are independent of functions of past elements of u_t. In particular we have

$$E(X_{i_1\,t-1}^{p_1} \cdot X_{i_2\,t-1}^{p_2} \cdot \dots \cdot X_{i_g\,t-1}^{p_g} \cdot u_{j_1 t}^{v_1} \cdot u_{j_2 t}^{v_2} \cdot \dots \cdot u_{j_h t}^{v_h}) \tag{E.5}$$

$$= E(X_{i_1\,t-1}^{p_1} \cdot X_{i_2\,t-1}^{p_2} \cdot \dots \cdot X_{i_g\,t-1}^{p_g}) \cdot E(u_{j_1 t}^{v_1} \cdot u_{j_2 t}^{v_2} \cdot \dots \cdot u_{j_h t}^{v_h}) \tag{E.6}$$

for all t, g, h,$i_1, \dots, i_g$, $j_1, \dots, j_h$, $p_1, \dots, p_g$, $v_1, \dots, v_g$.

E.2 Moments of Factors

Let X_{it} denote the ith element of the vector X_t. The process $\{X_t\}$ has expectation zero,

$$E(X_{it}) = 0. \tag{E.7}$$

Second moments are given as

$$\begin{aligned} & E(X_{it} X_{jt}) \\ &= E[(\kappa_i X_{i\,t-1} + u_{it})(\kappa_j X_{j\,t-1} + u_{jt})] \\ &= \kappa_i \kappa_j E(X_{it} X_{jt}) + E(u_{it} u_{jt}) \end{aligned} \tag{E.8}$$

The last equation holds since cross product terms vanish by (E.5) and E.1). Moreover, $E(X_{i\,t-1}X_{j\,t-1}) = E(X_{it}X_{jt})$ by stationarity.

For $i \neq j$, the last term vanishes by (E.2), since $\mu_b = 0$ is assumed, for $i = j$ it does not. Using stationarity we have

$$E(X_{it}^2) = \frac{E(u_{it}^2)}{1-\kappa_i^2}, \tag{E.9}$$

and

$$E(X_{it}X_{jt}) = 0 \text{ for } i \neq j \tag{E.10}$$

For third moments we have

$$\begin{aligned} &E(X_{it}X_{jt}X_{kt}) \\ &= E[(\kappa_i X_{i\,t-1} + u_{it})(\kappa_j X_{j\,t-1} + u_{jt})(\kappa_k X_{k\,t-1} + u_{kt})] \\ &= 0 \end{aligned} \tag{E.11}$$

Expanding the product of the second equality, one is left with three types of terms. First, we have expressions like $E(X_{i\,t-1}X_{j\,t-1}u_{kt})$ and $E(X_{i\,t-1}u_{jt}u_{kt})$. By (E.5) the first term equals $E(X_{i\,t-1}X_{j\,t-1}) \cdot E(u_{kt})$, which is equal to zero since the second factor is zero. Similarly, $E(X_{i\,t-1}u_{jt}u_{kt}) = E(X_{i\,t-1}) \cdot E(u_{jt}u_{kt})$ of which the first factor is zero. Second, we have the term $E(u_{it} \cdot u_{jt} \cdot u_{kt})$. If i, j and k are all different, we apply (E.3), if exactly two of them are equal we use (E.2), and for $i = j = k$ (E.1) is employed. In all cases the expression is zero, which follows from the assumption $\mu_b = 0$. Third, we have $\kappa_i\kappa_j\kappa_k E(X_{i\,t-1}X_{j\,t-1}X_{k\,t-1})$, which equals $\kappa_i\kappa_j\kappa_k E(X_{it}X_{jt}X_{kt})$ by stationarity. So one is left with $E(X_{it}X_{jt}X_{kt}) = \kappa_i\kappa_j\kappa_k E(X_{it}X_{jt}X_{kt})$, which implies that the third moment is zero.

Similar arguments hold for the computation of fourth moments.

$$\begin{aligned} &E(X_{it}X_{jt}X_{kt}X_{lt}) \\ &= E[(\kappa_i X_{i\,t-1} + u_{it})(\kappa_j X_{j\,t-1} + u_{jt})(\kappa_k X_{k\,t-1} + u_{kt})(\kappa_l X_{l\,t-1} + u_{lt})] \end{aligned}$$

Expanding the right hand side, terms of the form $E(X_{i\,t-1}u_{jt}u_{kt}u_{lt})$ and of the form $E(X_{i\,t-1}X_{j\,t-1}X_{k\,t-1}u_{lt})$ are zero. The remaining terms are given as as follows:

$$\begin{aligned} &E(X_{it}X_{jt}X_{kt}X_{lt}) \\ &= \kappa_i\kappa_j\kappa_k\kappa_l E(X_{it}X_{jt}X_{kt}X_{lt}) \\ &\quad +\kappa_i\kappa_j E(X_{it}X_{jt})E(u_{kt}u_{lt}) + \kappa_i\kappa_k E(X_{it}X_{kt})E(u_{jt}u_{lt}) \\ &\quad +\kappa_i\kappa_l E(X_{it}X_{lt})E(u_{jt}u_{kt}) + \kappa_j\kappa_k E(X_{jt}X_{kt})E(u_{it}u_{lt}) \\ &\quad +\kappa_j\kappa_l E(X_{jt}X_{lt})E(u_{it}u_{kt}) + \kappa_k\kappa_l E(X_{kt}X_{lt})E(u_{it}u_{jt}) \\ &\quad +E(u_{it}u_{jt}u_{kt}u_{lt}) \end{aligned}$$

This expression can be simplified. The result depends on which of the indices show up repeatedly. Using (E.2) and (E.4), one obtains that

$$E(X_{it}^3 X_{jt}) = 0, \quad i \neq j, \tag{E.12}$$
$$E(X_{it}^2 X_{jt} X_{kt}) = 0, \quad i,\ j,\ k \text{ are all different}, \tag{E.13}$$
$$E(X_{it} X_{jt} X_{kt} X_{lt}) = 0, \quad i,\ j,\ k,\ l \text{ are all different}. \tag{E.14}$$

Positive expressions only result from index combinations of the form $E(X_{it}^4)$ and $E(X_{it}^2 X_{jt}^2)$. We have

$$E(X_{it}^4) = \frac{6\kappa_i^2 E(X_{it}^2)E(u_{it}^2) + E(u_{it}^4)}{1-\kappa_i^4}, \tag{E.15}$$
$$E(X_{it}^2 X_{jt}^2) = \frac{\kappa_i^2 E(X_{it}^2)E(u_{jt}^2) + \kappa_j^2 E(X_{jt}^2)E(u_{it}^2) + E(u_{it}^2 u_{jt}^2)}{1-\kappa_i^2\kappa_j^2}. \tag{E.16}$$

Finally, we compute the autocovariance of the factor processes. Since each individual factor follows an AR(1) process, we have that

$$E(X_{it} X_{i\,t-h}) = \kappa_i^h E(X_{it}^2). \tag{E.17}$$

E.3 Moments of Differenced Factors

Moments of first differences can be easily expressed in terms of the moments of levels that have just been computed. We have

$$\begin{aligned} \Delta X_{it} &= X_{it} - X_{i\,t-1} \\ &= \kappa_i X_{i\,t-1} + u_{it} - X_{i\,t-1} \\ &= -(1-\kappa_i)X_{i\,t-1} + u_{it}. \end{aligned}$$

Obviously,

$$E(\Delta X_t) = 0. \tag{E.18}$$

For second moments we have

$$\begin{aligned} &E(\Delta X_{it}\, \Delta X_{jt}) \\ &= E[(-(1-\kappa_i)X_{i\,t-1} + u_{it})(-(1-\kappa_j)X_{j\,t-1} + u_{jt})] \\ &= (1-\kappa_i)(1-\kappa_j)E(X_{it}X_{jt}) + E(u_{it}u_{jt}) \end{aligned} \tag{E.19}$$

Note that the last line is the same as in (E.8) except that κ_i and κ_j are replaced by $1-\kappa_i$ and $1-\kappa_j$ respectively. Again, second cross moments vanish,

$$E(\Delta X_{it} \Delta X_{jt}) = 0, \quad i \neq j \tag{E.20}$$

The variance of ΔX_{it}, expressed in terms of the variance of X_{it} is given as

$$E(\Delta X_{it}{}^2) = (1-\kappa_i)^2 E(X_{it}^2) + E(u_{it}^2). \tag{E.21}$$

Similar arguments lead to the formulas for third and fourth moments of first differences expressed in the corresponding moments of the level variables. For third moments,

$$E(\Delta X_{it}\Delta X_{jt}\Delta X_{kt}) = 0. \tag{E.22}$$

For fourth moments:

$$\begin{aligned} &E(\Delta X_{it}{}^4) \\ &= (1-\kappa_i)^4 E(X_{it}^4) + 6(1-\kappa_i)^2 E(X_{it}^2)E(u_{it}^2) + E(u_{1t}^4) \end{aligned} \tag{E.23}$$

$$\begin{aligned} &E(\Delta X_{it}{}^2 \Delta X_{jt}{}^2) \\ &= (1-\kappa_i)^2(1-\kappa_j)^2 E(X_{it}^2 X_{jt}^2) \\ &\quad +(1-\kappa_i)^2 E(X_{it}^2)E(u_{jt}^2) + (1-\kappa_j)^2 E(X_{jt}^2)E(u_{it}^2) \\ &\quad +E(u_{it}^2 u_{jt}^2) \end{aligned} \tag{E.24}$$

$$E(\Delta X_{it}{}^3 \Delta X_{jt}) = 0, \quad i \neq j, \tag{E.25}$$

$$E(\Delta X_{it}{}^2 \Delta X_{jt}\Delta X_{kt}) = 0, \quad i,\ j,\ k \text{ are all different}, \tag{E.26}$$

$$E(\Delta X_{it}\Delta X_{jt}\Delta X_{kt}\Delta X_{lt}) = 0, \quad i,\ j,\ k,\ l \text{ are all different}. \tag{E.27}$$

For first differences, the autocovariance is given by

$$\begin{aligned} E(\Delta X_{it}\cdot \Delta X_{i\,t-h}) &= \kappa_i^h E(X_{it}^2) \\ &= E[(X_{it} - X_{i\,t-1})(X_{i\,t-h} - X_{i\,t-h-1})] \\ &= E(X_{it}^2)\cdot(\kappa_i^h - \kappa_i^{h+1} - \kappa_i^{h-1} + \kappa_i^h) \\ &= E(X_{it}^2)\kappa_i^h\left(1 - \kappa_i - \frac{1}{\kappa_i} + 1\right) \\ &= E(X_{it}^2)\kappa_i^h\left(-\frac{\kappa_i^2 - 2\kappa_i + 1}{\kappa_i}\right) \\ &= -E(X_{it}^2)\kappa_i^{h-1}(1-\kappa_i)^2 \end{aligned} \tag{E.28}$$

E.4 Moments of Differenced Yields

Yields in first differences depend on differenced factors in the same way as yields in levels depend on factors in levels, we have

$$\begin{aligned} \Delta y_t^n &= y_t^n - y_{t-1}^n \\ &= \frac{A_n}{n} + \sum_{i=1}^d \frac{B_{in}}{n} X_{it} - \frac{A_n}{n} - \sum_{i=1}^d \frac{B_{in}}{n} X_{i\,t-1} \\ &= \sum_{i=1}^d \frac{B_{in}}{n}\Delta X_{it} \end{aligned}$$

First differences have mean zero,

$$E[\Delta y_t^n] = 0, \tag{E.29}$$

the variance becomes

$$E[(\Delta y_t^n)^2] = \sum_{i=1}^{d} \left(\frac{B_{in}}{n}\right)^2 E(\Delta X_{it}{}^2). \tag{E.30}$$

It is straightforward to show that $\kappa_i > 0.5$ for all i is a sufficient condition for the variance of yields in levels to exceed the variance of yield changes.

The covariance is given as

$$E[(\Delta y_t^n)(\Delta y_t^m)] = \sum_{i=1}^{d} \frac{B_{in}}{n}\frac{B_{im}}{m} E(\Delta X_{it}{}^2), \tag{E.31}$$

thus the contemporaneous correlation between two different yield differences is the same as for yields in levels,

$$Corr(\Delta y_t^n, \Delta y_t^m) = \frac{\sum_{i=1}^{d} B_{in} B_{im} E(\Delta X_{it}{}^2)}{\sqrt{\sum_{i=1}^{d} B_{in}^2\, E(\Delta X_{it}{}^2)} \cdot \sqrt{\sum_{i=1}^{d} B_{im}^2\, E(\Delta X_{it}{}^2)}}. \tag{E.32}$$

Third moments are zero,

$$E[(\Delta y_t^n)^3] = 0 \tag{E.33}$$

and fourth moments turn out to be

$$E[(\Delta y_t^{*n})^4] = \sum_{i=1}^{d}\sum_{j=1}^{d} \left(\frac{B_{in}}{n}\right)^2 \left(\frac{B_{jn}}{n}\right)^2 E(\Delta X_{it}{}^2 \Delta X_{jt}{}^2). \tag{E.34}$$

For fourth moments we have a similar decomposition as for yields in levels. We can write

$$E(\Delta X_{it}{}^2 \Delta X_{jt}{}^2) = E(\Delta X_{it}{}^2 \Delta X_{jt}{}^2)_{Gauss} + w_{ij}^{\Delta} \cdot d_{ij} \tag{E.35}$$

with

$$w_{ij}^{\Delta} = \frac{(1-\kappa_i)^2(1-\kappa_j)^2}{1-\kappa_i^2\kappa_j^2} + 1$$

and d_{ij} defined by (3.111). Thus,

$$\begin{aligned} E[(\Delta y_t^n)^4] &= \sum_{i=1}^{d}\sum_{j=1}^{d} \left(\frac{B_{in}}{n}\right)^2 \left(\frac{B_{jn}}{n}\right)^2 \\ &= E[(\Delta y_t^n)^4]_{Gauss} + \sum_{i=1}^{d}\sum_{j=1}^{d} \left(\frac{B_{in}}{n}\right)^2 \left(\frac{B_{jn}}{n}\right)^2 w_{ij}^{\Delta} \cdot d_{ij}, \end{aligned}$$

and for the kurtosis of yield changes

$$kurt(\Delta y_t^n) = \frac{\sum_{i=1}^d \sum_{j=1}^d \left(\frac{B_{in}}{n}\right)^2 \left(\frac{B_{jn}}{n}\right)^2 w_{ij}^{\Delta} \cdot d_{ij}}{(E[(\Delta y_t^{n*})^2])^2} \tag{E.36}$$

Finally, the autocorrelation structure of yields and yield changes is given as follows. For the factors we have

$$E(X_{it} X_{i\,t-h}) = \kappa_i^h E(X_{it}^2) \tag{E.37}$$

and

$$E(\Delta X_{it} \cdot \Delta X_{i\,t-h}) = \kappa_i^{h-1}(1-\kappa_i)^2 E(X_{it}^2). \tag{E.38}$$

Any two different factors are uncorrelated over time,

$$E(X_{it} X_{j\,t-h}) = E(\Delta X_{it} \cdot \Delta X_{j\,t-h}) = 0. \tag{E.39}$$

Thus, for the autocorrelation function of yields it follows that

$$Corr(y_t^n, y_{t-h}^n) = \frac{\sum_{i=1}^d \left(\frac{B_{in}}{n}\right)^2 \kappa_i^h E(X_{it}^2)}{\sum_{i=1}^d \left(\frac{B_{in}}{n}\right)^2 E(X_{it}^2)} \tag{E.40}$$

and for yield changes we have

$$Corr(\Delta y_t^n, \Delta y_{t-h}^n) = -\frac{\sum_{i=1}^d \left(\frac{B_{in}}{n}\right)^2 \kappa_i^{h-1}(1-\kappa_i)^2 E(X_{it}^2)}{\sum_{i=1}^d \left(\frac{B_{in}}{n}\right)^2 2(1-\kappa_i) E(X_{it}^2)}. \tag{E.41}$$

F

Proof of Theorem 6.3

Denote by $p_2(x) = \phi(x;\ \mu, V)$ an arbitrary k-variate simple normal density. The Kullback-Leibler distance of $p_2(x)$ to $p_1(x)$ is given by

$$KL(p_1, p_2) = \int \ln \frac{p_1(x)}{p_2(x)} p_1(x) dx.$$

We interpret this distance as a function of μ and V and define $G(\mu, V) := KL(p_1, p_2)$. We will show that μ_c and V_c in (6.34) satisfy the first order conditions for minimizers of $G(\mu, V)$. For notational convenience we may drop the arguments of p_1, p_2. Moreover, we use the short hand notation ϕ_i for the component density $\phi(x;\ \mu_i, V_i)$. We have

$$\begin{aligned} G(\mu, V) &= \int \ln \frac{p_1}{p_2} \left(\sum_i \omega_i \phi_i \right) dx \\ &= \sum_i \omega_i \int \ln p_1\, \phi_i dx - \sum_i \omega_i \int \ln p_2\, \phi_i\, dx. \end{aligned}$$

The first order conditions are given by

$$\frac{\partial G(\mu, V)}{\partial \mu} = -\sum_i \omega_i \int \frac{\partial \ln p_2}{\partial \mu} \phi_i\, dx = 0 \tag{F.1}$$

and

$$\frac{\partial G(\mu, V)}{\partial V} = -\sum_i \omega_i \int \frac{\partial \ln p_2}{\partial V} \phi_i\, dx = 0. \tag{F.2}$$

Next, we compute the derivatives in these two equations. First note that

$$\ln p_2 = -\ln \left[(2\pi)^{k/2} \sqrt{|V_c|} \right] - \frac{1}{2}(x - \mu_c)' V_c^{-1} (x - \mu_c). \tag{F.3}$$

We have

$$\frac{\partial \ln p_2}{\partial \mu} = (x - \mu_c)' V_c^{-1} \tag{F.4}$$

and

$$\begin{aligned}
&\frac{\partial \ln p_2}{\partial V} \\
&= -\frac{1}{(2\pi)^{k/2}\sqrt{|V_c|}} \cdot (2\pi)^{k/2} \cdot \frac{1}{2\sqrt{|V_c|}} \cdot \frac{\partial |V_c|}{\partial V_c} \\
&\quad + \frac{\partial \left[-\frac{1}{2}(x - \mu_c)' V_c^{-1} (x - \mu_c)\right]}{\partial V_c} \\
&= -\frac{1}{2} V_c^{-1} + \frac{1}{2} V_c^{-1} (x - \mu_c)(x - \mu_c)' V_c^{-1}.
\end{aligned} \tag{F.5}$$

For the last equality it has been used that for a symmetric matrix A and a vector b,

$$\begin{aligned}
\frac{\partial |A|}{\partial A} &= |A| \cdot A^{-1}, \\
\frac{\partial\, b' A^{-1} b}{\partial A} &= -A^{-1} b\, b' A^{-1},
\end{aligned}$$

see [77], pages 181 and 177, respectively

Inserting the derivative (F.4) into the first order condition (F.1) and transforming to a column vector yields

$$\sum_i \omega_i \int V_c^{-1} (x - \mu_c) \phi_i \, dx = 0. \tag{F.6}$$

Computing the left hand side leads to

$$\begin{aligned}
&V_c^{-1} \sum_i \omega_i \int (x - \mu_c) \phi_i \, dx \\
&= V_c^{-1} \sum_i \omega_i \left(\int x \phi_i \, dx - \int \mu_c \phi_i \, dx \right) \\
&= V_c^{-1} \sum_i \omega_i \left(\mu_i - \mu_c \right).
\end{aligned}$$

Thus, (F.6) implies

$$\mu_c = \sum_i \omega_i \mu_i. \tag{F.7}$$

With (F.5), the second first order condition (F.2) becomes

$$\sum_i \omega_i \int \left(-\frac{1}{2} V_c^{-1} + \frac{1}{2} V - c^{-1} (x - \mu_c)(x - \mu_c)' V_c^{-1} \right) \phi_i \, dx = 0. \tag{F.8}$$

The second integral involved is computed as

$$\begin{aligned}
&\int (x-\mu_c)(x-\mu_c)'\phi_i\,dx \\
&= \int (x-\mu_i+\mu_i-\mu_c)(x-\mu_i+\mu_i-\mu_c)'\phi_i\,dx \\
&= \int \left[(x-\mu_i)(x-\mu_i)' + (x-\mu_i)(\mu_i-\mu_c)' \right. \\
&\quad \left. +(\mu_i-\mu_c)(x-\mu_i)' + (\mu_i-\mu_c)(\mu_i-\mu_c)'\right]\phi_i\,dx \\
&= V_i + 0 + 0 + (\mu_i-\mu_c)(\mu_i-\mu_c)'.
\end{aligned}$$

Thus, (F.8) becomes

$$-\frac{1}{2}V_c^{-1} + \frac{1}{2}V_c^{-1}\left(\sum_i \omega_i\left[V_i + (\mu_i-\mu_c)(\mu_i-\mu_c)'\right]\right)V_c^{-1} = 0. \tag{F.9}$$

Multiplying through by $-2V_c$ from the right and by V_c from the left yields

$$V_c - \sum_i \omega_i\left[V_i + (\mu_i-\mu_c)(\mu_i-\mu_c)'\right] = 0, \tag{F.10}$$

which completes the proof.

G

Random Draws from a Gaussian Mixture Distribution

For our simulation study in chapter 7 we need to draw random variates from the normal distribution and the Gaussian mixture distribution. We now show how draws from these distributions can be obtained by transforming random variates from the uniform distribution and the univariate standard normal distribution. This is done for the general case of g-dimensional random vectors and B components in the mixture distribution.

For a draw from the g-variate normal distribution with mean vector μ and variance-covariance matrix Q, one first generates a vector Z of g independent random variates from $N(0,1)$. The transformed variable X with

$$X = \mu + CZ$$

where $CC' = Q$ is the Choleski decomposition of Q, can then be treated as a draw from $N(\mu, Q)$.

Consider now the problem of drawing at random from the g-dimensional multivariate normal mixture with B components,

$$\sum_{b=1}^{B} \omega_b \, N(\mu_b, \, Q_b).$$

Generating pseudo random variables from this distribution is accomplished by a two step approach: first, a component j is drawn from $\{1, \ldots, B\}$ according to the probabilities $\omega_1, \ldots, \omega_B$. Second, a random draw from $N(\mu_j, \, Q_j)$ is made according to the procedure described above. In the following we present the algorithm that we use for randomly drawing the index j from $\{1, \ldots, B\}$.

Let v be the cumulative sum vector constructed from the probabilities $\omega_1, \ldots, \omega_B$ whose ith entry is given by

$$v_i = \begin{cases} 0 & i = 1 \\ \sum_{k=1}^{i-1} \omega_k & i = 2, \ldots, B \end{cases} .$$

Draw U from $\mathcal{U}(0,1)$ and construct the $B \times 1$ vector h whose components are given by

$$h_i = \begin{cases} 1 & U > v_i \\ 0 & \text{else} \end{cases} \qquad i = 1, \ldots, B.$$

Compute j as

$$j = h_1 + \ldots h_B.$$

Now we have to show that this procedure guarantees that the component indices of the mixture are chosen with the correct probabilities. That is, we have to show that for any $l \in \{1, \ldots, B\}$, $P(j = l) = \omega_l$.

We have $P(j = l) = P(h_1 + \ldots + h_B = l)$. Now, $h_1 + \ldots + h_B = l$ if and only if[1]

$$U > \sum_{k=1}^{l-1} \omega_k \quad \text{and} \quad U \leq \sum_{k=1}^{l} \omega_k$$

which follows from the definition of the vector h. Thus,

$$P(j = l) = P\left(U \in \left(\sum_{k=1}^{l-1} \omega_k, \sum_{k=1}^{l} \omega_k\right]\right) = \sum_{k=1}^{l} \omega_k - \sum_{k=1}^{l-1} \omega_k = \omega_l,$$

which proves the assertion.

[1] For $l = 1$ the empty sum is zero.

References

1. Ackerson GA, Fu KS (1970) On State Estimation in Switching Environments. IEEE Transactions on Automatic Control 15:10–17
2. Akashi H, Kumamoto H (1977) Random Sampling Approach to State Estimation in Switching Environments. Automatica 13:429–434
3. Alspach DL, Sorenson HW (1972) Nonlinear Bayesian Estimation Using Gaussian Sum Approximations. IEEE Transactions on Automatic Control 17:439–448
4. Anderson BDO, Moore JB (1979) Optimal Filtering. Prentice Hall, Englewood Cliffs NJ
5. Anderson N, Breedon F, Deacon M, Derry A, Murphy G (1996) Estimating and Interpreting the Yield Curve. Wiley, Chichester et al
6. Ang A, Piazzesi M (2003) A No-Arbitrage Vector Autoregression of Term Structure Dynamics with Macroeconomic and Latent Variables. Journal of Monetary Economics 50:745-787
7. Aoki M (1990) State Space Modeling of Time Series. Springer, Berlin et al, 2nd edition
8. Baadsgaard M, Nielsen JN, Madsen H (2000) Estimating Multivariate Exponential-Affine Term Structure Models from Coupon Bond Prices using Nonlinear Filtering. IMM Technical Report 09-2000, Technical University of Denmark
9. Babbs SH, Nowman KB (1998) An Application of Generalized Vasicek Term Structure Models to the UK Gilt-edged Market: a Kalman Filter Analysis. Applied Financial Economics 8:637–644
10. Babbs SH, Nowman KB (1999) Kalman Filtering of Generalized Vasicek Term Structure Models. Journal of Financial and Quantitative Analysis 34:115–130
11. Backus D, Foresi S, Telmer C (1998) Discrete-Time Models of Bond Pricing. NBER Working Paper 6736
12. Ball CA, Torous WN (1996) Unit Roots and the Estimation of Interest Rate Dynamics. Journal of Empirical Finance 3:215–238
13. Bansal R, Zhou H (2002) Term Structure of Interest Rates with Regime Shifts. Journal of Finance 57:1997–2044
14. Baxter M, Rennie A (1999) Financial Calculus. Cambridge University Press, Cambridge et al

15. Beaglehole D, Tenney M (1992) Corrections and Additions to "A Nonlinear Equilibrium Model of the Term Structure of Interest Rates".Journal of Financial Economics 32:345–353
16. Bhar R, Chiarella C (1997) Estimation of the Heath-Jarrow-Morton Model by Use of Kalman Filtering Techniques. In: Amman H, Rustem B, Whinston A (eds) Computational Approaches to Economic Problems. Kluwer, Dordrecht et al.
17. Bingham NH, Kiesel R (1998) Risk Neutral Valuation. Springer, London et al
18. Björk T (1996) Interest rate theory. In: Runggaldier WJ (ed) Financial Mathematics - Lectures given at the 3rd Session of the Centro Internazionale Matematico Estivo held in Bressanone, Italy, July 8-13, 1996. Springer, Berlin et al
19. Björk T (1998) Arbitrage Theory in Continous Time. Oxford University Press, Oxford et al
20. Bliss RR (1997) Testing Term Structure Estimation Methods. Advances in Futures and Options Research 9:197–231
21. Bolstad WM (1995) The Multiprocess Dynamic Poisson Model. Journal of the American Statistical Association 90:227–2382
22. Brigo D, Hanzon B (1998) On some Filtering Problems Arising in Mathematical Finance. Insurance: Mathematics and Economics 22:53–64
23. Brockwell PJ, Davis RA (1991) Time Series : Theory and Methods. Springer, New York et al, 2nd edition
24. Brockwell PJ, Davis RA (1996) Introduction to Time Series and Forecasting. Springer, New York et al
25. Brown RH, Schaefer SM (1994) The Term Structure Model of Real Interest Rates and the Cox, Ingersoll, and Ross Model. Journal of Financial Economics 35:3–42
26. Campbell JY, Lo AW, MacKinlay A (1997) The Econometrics of Financial Markets. Princeton University Press, Princeton
27. Campbell JY (2000) Asset Pricing at the Millenium. The Journal of Finance 55:1515–1567
28. Cassola N, Luis JB (2003) A Two-Factor Model of the German Term Structure of Interest Rates. Applied Financial Economics 13:783–806
29. Chan K, Karoly G, Longstaff F, Sanders A (1992) An Empirical Comparison of Alternative Models of the Short-Term Interest Rate. Journal of Finance 47:1209–1227
30. Chen L (1996) Interest Rate Dynamics, Derivatives Pricing, and Risk Management. Springer, Heidelberg et al
31. Chen R, Liu JS (2000) Mixture Kalman Filters. Journal of the Royal Statistical Society, Series B 62:493–508
32. Chow HK (1994) Robust Estimation in Time Series: An Approximation to the Gaussian Sum Filter. Communications in Statistics - Theory and Methods 23:3491–3505
33. Cochrane J (2001) Asset Pricing. Princeton University Press, Princeton et al
34. Cox JC, Ingersoll JE, Ross SA (1985) A Theory of the Term Structure of Interest Rates. Econometrica 53:385–407
35. Dai Q, Singleton KJ (2000) Specification Analysis of Affine Term Structure Models. The Journal of Finance 55:1943–1978
36. Dai Q, Singleton KJ, Yang W (2003) Regime Shifts in a Dynamic Term Structure Model of U.S. Treasury Bond Yields. Working Paper Fin-03-040, New York University - Stern School of Business

37. Davidson R, MacKinnon JG (1993) Estimation and Inference in Econometrics. Oxford University Press, New York et al
38. de Jong F (2000) Time Series and Cross-Section Information in Affine Term-Structure Models. Journal of Business and Economic Statistics 18:300–314
39. de Jong F, Santa-Clara P (1999) The Dynamics of the Forward Interest Rate Curve: A Formulation with State Variables. Journal of Financial and Quantitative Analysis 34:131–157
40. Duan J, Simonato J (1999) Estimating and Testing Exponential-Affine Term Structure Models by Kalman Filter. Review of Quantitative Finance and Accounting 13:111–135
41. Duffee GR (1999) Estimating the Price of Default Risk. The Review of Financial Studies 12:197–226
42. Duffee GR (2002) Term premia and Interest Rate Forecasts in Affine Models. Journal of Finance 57:405–443
43. Duffie D (1996a) Dynamic Asset Pricing Theory. Princeton University Press, Princeton NJ, 2nd edition
44. Duffie D (1996b) State Space Models of the Term Structure of Interest Rates. In: Hughston L (ed) Vasicek and Beyond. Risk Publications, London
45. Duffie D, Kan R (1996) A Yield-Factor Model of Interest Rates. Mathematical Finance, 6:379–406
46. Durbin J, Koopman SJ (2001) Time Series Analysis by State Space Methods. Oxford University Press, Oxford et al
47. Efron B, Tibshirani RJ (1993) An Introduction to the Bootstrap. Chapman and Hall, New York et al
48. Fahrmeir L, Künstler R, Pigeot I, Tutz G (1997) Statistik. Springer, Berlin et al
49. Fletcher R (1987) Practical Methods of Optimization. Wiley, Chichester et al., 2nd edition
50. Frühwirth R (1995) Track Fitting with Long-Tailed Noise: A Bayesian Approach. Computer Physics Communications 85:189–199
51. Frühwirth R (1997) Track Fitting with Non-Gaussian Noise. Computer Physics Communications 100:1–16
52. Frühwirth-Schnatter S, Geyer AL (1996) Bayesian Estimation of Econometric Multi-Factor Cox-Ingersoll-Ross-Models of the Term Structure of Interest Rates Via MCMC Methods. Working Paper, Vienna University of Economics and Business Administration
53. Fuller WA (1996) Introduction to Statistical Time Series. Wiley, New York et al, 2nd edition
54. Geweke J, Tanizaki H (2001) Bayesian Estimation of State-Space Models Using the Metropolis-Hastings Algorithm within Gibbs Sampling. Computational Statistics and Data Analysis 37:151–170
55. Geyer AL, Pichler S (1999) A State Space Approach to Estimate and Test Multifactor Cox-Ingersoll-Ross Models of the Term Structure Journal of Financial Research 22:107–130
56. Gourieroux C, Monfort A (1997) Time Series and Dynamic Models. Cambridge University Press, Cambridge et al
57. Hamilton J (1994) Time Series Analysis. Princeton University Press, Princeton NJ
58. Harrison PJ, Stevens CF (1976) Bayesian Forecasting. Journal of the Royal Statistical Society, Series B 38:205-247

59. Harvey A (1990) Forecasting, Structural Time Series Models and the Kalman Filter. Cambridge University Press, Cambridge et al
60. Harvey A (1993) Time Series Models. Harvester Wheatsheaf, New York et al, 2nd edition
61. Heath D, Jarrow R, Morton A (1992) Bond Pricing and the Term Structure of Interest Rates: A New Methodology for Contingent Claims Valuation. Econometrica 60:77–105
62. Hördahl P, Tristani O, Vestin D (2005) A Joint Econometric Model of Macroeconomic and Term-Structure Dynamics. Journal of Econometrics, forthcoming
63. Ingersoll JE (1987) Theory of Financial Decision Making. Rowman and Littlefield, Totowa NJ
64. Irle A (1998) Finanzmathematik - Die Bewertung von Derivaten. Teubner, Stuttgart
65. James J, Webber N (2000) Interest Rate Modelling. Wiley, Chichester et al
66. Jegadeesh N, Pennacchi GG (1996) The Behavior of Interest Rates Implied by the Term Structure of Eurodollar Futures. Journal of Money, Credit, and Banking 28:426–451
67. Johnson NL, Kotz S (1970) Distributions in Statistics - Continuous Univariate Distributions 1. Houghton Mifflin, Boston et al
68. Kalman RE (1960) A New Approach to Linear Filtering and Prediction Problems. Transactions ASME, Journal of Basic Engineering 82:35–45
69. Kalman RE, Bucy RS (1961) New Results in Linear Filtering and Prediction Theory. Transactions ASME, Journal of Basic Engineering 83:95–103
70. Kellerhals BP (2001) Financial Pricing Models in Continuous Time and Kalman Filtering. Springer, Berlin et al
71. Kim CJ (1994) Dynamic linear models with Markov Switching. Journal of Econometrics 60:1–22
72. Kim CJ, Nelson CR (1999) State Space Models with Regime Switching - Classical and Gibbs-Sampling Approaches with Applications MIT Press, Cambridge et al
73. Kitagawa G (1989) Non-Gaussian Seasonal Adjustment. Computers and Mathematics with Applications 18:503–514
74. Klebaner FC (1998) Introduction to Stochastic Calculus with Applications. Imperial College Press, London
75. Lin DK, Guttman I (1993) Handling Spuriosity in the Kalman Filter. Statistics and Probability Letters 16:259–268
76. Lütkepohl H (1993) Introduction to Multiple Time Series Analysis. Springer, Berlin et al, 2nd edition
77. Lütkepohl H (1996) Handbook of Matrices. Wiley, Chichester et al
78. Lund J (1997a) Econometric Analysis of Continous-Time Arbitrage-Free Models of the Term Structure of Interest Rates. Working Paper, The Aarhus School of Business, Department of Finance
79. Lund J (1997b) Non-Linear Kalman Filtering Techniques for Term-Structure Models. Working Paper, The Aarhus School of Business, Department of Finance
80. Masreliez CJ (1975) Approximate Non-Gaussian Filtering with Linear State and Observation Relations. IEEE Transactions on Automatic Control 20:107–110
81. Maybeck PS (1982) Stochastic Models, Estimation, and Control, Vol. 2. Academic Press, New York et al

82. McCulloch JH (1971) Measuring the Term Structure of Interest Rates. The Journal of Business 44:19–31
83. McCulloch JH (1975) The Tax-Adjusted Yield Curve. Journal of Finance 30:811–830
84. McCulloch JH, Kwon HC (1993) U.S. Term Structure Data, 1947-1991. Working Paper 93-6, Ohio State University
85. McLachlan G, Peel D (2000) Finite Mixture Models. Wiley, New York et al
86. Mood AM, Graybill FA, Boes DC (1974) Introduction to the Theory of Statistics. McGraw-Hill, New York et al, 3rd edition
87. Musiela M, Rutkowski M (1997) Martingale Methods in Financial Modelling. Springer, Berlin et al
88. Nelson CR, Siegel AF (1987) Parsimonious Modeling of Yield Curves. The Journal of Business 60:473–489
89. Pagan AR, Hall AD, Martin V (1996) Modeling the Term Structure. In: Maddala GS, Rao CR (eds) Handbook of Statistics, Vol. 14. North-Holland, Amsterdam et al
90. Pena D, Guttman I (1989) Optimal Collapsing of Mixture Distributions in Robust Recursive Estimation. Communications in Statistics - Theory and Methods 18:817–833
91. Pennacchi GG (1991) Identifying the Dynamics of Real Interest Rates and Inflation: Evidence using Survey Data. The Review of Financial Studies 4:53–86
92. Piazzesi M (2005) Affine Term Structure Models. In: Ait-Sahalia Y, Hansen L (eds) Handbook of Financial Econometrics, forthcoming
93. Polson NG, Stroud JR, Müller P (2002) Affine State-Dependent Variance Models. mimeo
94. Rebonato R (1998) Interest Rate Option Models: Understanding, Analysing and Using Models for Exotic Interest-Rate Options. Wiley, Chichester et al, 2nd edition
95. Rudebusch G, Wu T (2004) A Macro-Finance Model of the Term Structure, Monetary Policy, and the Economy. Federal Reserve Bank of San Francisco Proceedings
96. Schmid B (2004) Credit Risk Pricing Models - Theory and Practice. Springer, Berlin et al, 2nd edition
97. Schwaar C (1999) Kalman-Filter basierte ML-Schätzung affiner, zeithomogener Faktormodelle der Zinsstruktur am bundesdeutschen Rentenmarkt. Europäische Hochschulschriften, Peter Lang, Frankfurt et al
98. Shephard N (1994) Partial Non-Gaussian State Space. Biometrica 81:115–131
99. Shiller R, McCulloch JH (1990) The Term Structure of Interest Rates. In: Friedman BM, Hahn FH (eds) Handbook of Monetary Economics, Vol. 1. North-Holland, Amsterdam et al
100. Shumway RH, Stoffer DS (2000) Time Series Analysis and Its Applications. Springer, New York et al
101. Singer H (1998) Finanzmarktökonometrie - Zeitstetige Systeme und Ihre Anwendung in Ökonometrie und Empirischer Kapitalmarktforschung. Physika-Verlag, Heidelberg
102. Sorenson HW, Alspach DL (1971) Recursive Bayesian Estimation Using Gaussian Sums. Automatica 7:465–479

103. Stroud JR, Müller P, Polson NG (2003) Nonlinear State-Space Models with State-Dependent Variance Functions. Journal of the American Statistical Association 98:377-386
104. Sun TS (1992) Real and Nominal Interest Rates: A Discrete-Time Model and its Continuous-Time Limit. Review of Financial Studies 5:581–611
105. Sundaresan SM (2000) Continuous-Time Methods in Finance: a Review and an Assessment. The Journal of Finance 55:1569–1622
106. Svensson LEO (1994) Estimating and Interpreting Forward Interest Rates: Sweden 1992 - 94. Working Paper 114, International Monetary Fund
107. Takahashi A, Sato S (2001) A Monte Carlo Filtering Approach for Estimating the Term Structure of Interest Rates. Annals of the Institute of Statistical Mathematics 53:50–62
108. Tanaka M, Katayama T (1987) Robust Kalman Filter for Linear Discrete-Time System with Gaussian Sum Noises. International Journal of Systems Science 18:1721–1731
109. Tanizaki H (1996) Nonlinear Filters - Estimation and Applications. Springer, Berlin et al, 2nd edition
110. Tanizaki H (2003) Nonlinear and Non-Gaussian State-Space Modeling with Monte Carlo Techniques: A Survey and Comparative Study. In: Shanbhag DN, Rao CR (eds) Handbook of Statistics, Vol. 21. North-Holland, Amsterdam et al
111. Titterington DM, Smith AFM, Makov UE (1985) Statistical Analysis of Finite Mixture Distributions. Wiley, Chichester et al
112. Tugnait JK (1982) Detection and Estimation of Abruptly Changing Systems. Automatica 18:607–615
113. Vasiček OA (1977) An Equilibrium Characterisation of the Term Structure. Journal of Financial Economics 5:177–188
114. Vasiček OA, Fong HG (1982) Term Structure Modeling Using Exponential Splines. The Journal of Finance 37:339–348
115. Williams D (2001) Weighing the Odds - A Course in Probability and Statistics. Cambridge University Press, Cambridge

List of Figures

List of Tables

Lecture Notes in Economics and Mathematical Systems

For information about Vols. 1–470
please contact your bookseller or Springer-Verlag

Vol. 471: N. H. M. Wilson (Ed.), Computer-Aided Transit Scheduling. XI, 444 pages. 1999.

Vol. 472: J.-R. Tyran, Money Illusion and Strategic Complementarity as Causes of Monetary Non-Neutrality. X, 228 pages. 1999.

Vol. 473: S. Helber, Performance Analysis of Flow Lines with Non-Linear Flow of Material. IX, 280 pages. 1999.

Vol. 474: U. Schwalbe, The Core of Economies with Asymmetric Information. IX, 141 pages. 1999.

Vol. 475: L. Kaas, Dynamic Macroeconomics with Imperfect Competition. XI, 155 pages. 1999.

Vol. 476: R. Demel, Fiscal Policy, Public Debt and the Term Structure of Interest Rates. X, 279 pages. 1999.

Vol. 477: M. Théra, R. Tichatschke (Eds.), Ill-posed Variational Problems and Regularization Techniques. VIII, 274 pages. 1999.

Vol. 478: S. Hartmann, Project Scheduling under Limited Resources. XII, 221 pages. 1999.

Vol. 479: L. v. Thadden, Money, Inflation, and Capital Formation. IX, 192 pages. 1999.

Vol. 480: M. Grazia Speranza, P. Stähly (Eds.), New Trends in Distribution Logistics. X, 336 pages. 1999.

Vol. 481: V. H. Nguyen, J. J. Strodiot, P. Tossings (Eds.). Optimation. IX, 498 pages. 2000.

Vol. 482: W. B. Zhang, A Theory of International Trade. XI, 192 pages. 2000.

Vol. 483: M. Königstein, Equity, Efficiency and Evolutionary Stability in Bargaining Games with Joint Production. XII, 197 pages. 2000.

Vol. 484: D. D. Gatti, M. Gallegati, A. Kirman, Interaction and Market Structure. VI, 298 pages. 2000.

Vol. 485: A. Garnaev, Search Games and Other Applications of Game Theory. VIII, 145 pages. 2000.

Vol. 486: M. Neugart, Nonlinear Labor Market Dynamics. X, 175 pages. 2000.

Vol. 487: Y. Y. Haimes, R. E. Steuer (Eds.), Research and Practice in Multiple Criteria Decision Making. XVII, 553 pages. 2000.

Vol. 488: B. Schmolck, Ommitted Variable Tests and Dynamic Specification. X, 144 pages. 2000.

Vol. 489: T. Steger, Transitional Dynamics and Economic Growth in Developing Countries. VIII, 151 pages. 2000.

Vol. 490: S. Minner, Strategic Safety Stocks in Supply Chains. XI, 214 pages. 2000.

Vol. 491: M. Ehrgott, Multicriteria Optimization. VIII, 242 pages. 2000.

Vol. 492: T. Phan Huy, Constraint Propagation in Flexible Manufacturing. IX, 258 pages. 2000.

Vol. 493: J. Zhu, Modular Pricing of Options. X, 170 pages. 2000.

Vol. 494: D. Franzen, Design of Master Agreements for OTC Derivatives. VIII, 175 pages. 2001.

Vol. 495: I. Konnov, Combined Relaxation Methods for Variational Inequalities. XI, 181 pages. 2001.

Vol. 496: P. Weiß, Unemployment in Open Economies. XII, 226 pages. 2001.

Vol. 497: J. Inkmann, Conditional Moment Estimation of Nonlinear Equation Systems. VIII, 214 pages. 2001.

Vol. 498: M. Reutter, A Macroeconomic Model of West German Unemployment. X, 125 pages. 2001.

Vol. 499: A. Casajus, Focal Points in Framed Games. XI, 131 pages. 2001.

Vol. 500: F. Nardini, Technical Progress and Economic Growth. XVII, 191 pages. 2001.

Vol. 501: M. Fleischmann, Quantitative Models for Reverse Logistics. XI, 181 pages. 2001.

Vol. 502: N. Hadjisavvas, J. E. Martínez-Legaz, J.-P. Penot (Eds.), Generalized Convexity and Generalized Monotonicity. IX, 410 pages. 2001.

Vol. 503: A. Kirman, J.-B. Zimmermann (Eds.), Economics with Heterogenous Interacting Agents. VII, 343 pages. 2001.

Vol. 504: P.-Y. Moix (Ed.),The Measurement of Market Risk. XI, 272 pages. 2001.

Vol. 505: S. Voß, J. R. Daduna (Eds.), Computer-Aided Scheduling of Public Transport. XI, 466 pages. 2001.

Vol. 506: B. P. Kellerhals, Financial Pricing Models in Con-tinuous Time and Kalman Filtering. XIV, 247 pages. 2001.

Vol. 507: M. Koksalan, S. Zionts, Multiple Criteria Decision Making in the New Millenium. XII, 481 pages. 2001.

Vol. 508: K. Neumann, C. Schwindt, J. Zimmermann, Project Scheduling with Time Windows and Scarce Resources. XI, 335 pages. 2002.

Vol. 509: D. Hornung, Investment, R&D, and Long-Run Growth. XVI, 194 pages. 2002.

Vol. 510: A. S. Tangian, Constructing and Applying Objective Functions. XII, 582 pages. 2002.

Vol. 511: M. Külpmann, Stock Market Overreaction and Fundamental Valuation. IX, 198 pages. 2002.

Vol. 512: W.-B. Zhang, An Economic Theory of Cities.XI, 220 pages. 2002.

Vol. 513: K. Marti, Stochastic Optimization Techniques. VIII, 364 pages. 2002.

Vol. 514: S. Wang, Y. Xia, Portfolio and Asset Pricing. XII, 200 pages. 2002.

Vol. 515: G. Heisig, Planning Stability in Material Requirements Planning System. XII, 264 pages. 2002.

Vol. 516: B. Schmid, Pricing Credit Linked Financial Instruments. X, 246 pages. 2002.

Vol. 517: H. I. Meinhardt, Cooperative Decision Making in Common Pool Situations. VIII, 205 pages. 2002.

Vol. 518: S. Napel, Bilateral Bargaining. VIII, 188 pages. 2002.

Vol. 519: A. Klose, G. Speranza, L. N. Van Wassenhove (Eds.), Quantitative Approaches to Distribution Logistics and Supply Chain Management. XIII, 421 pages. 2002.

Vol. 520: B. Glaser, Efficiency versus Sustainability in Dynamic Decision Making. IX, 252 pages. 2002.

Vol. 521: R. Cowan, N. Jonard (Eds.), Heterogenous Agents, Interactions and Economic Performance. XIV, 339 pages. 2003.

Vol. 522: C. Neff, Corporate Finance, Innovation, and Strategic Competition. IX, 218 pages. 2003.

Vol. 523: W.-B. Zhang, A Theory of Interregional Dynamics. XI, 231 pages. 2003.

Vol. 524: M. Frölich, Programme Evaluation and Treatment Choise. VIII, 191 pages. 2003.

Vol. 525: S. Spinler, Capacity Reservation for Capital-Intensive Technologies. XVI, 139 pages. 2003.

Vol. 526: C. F. Daganzo, A Theory of Supply Chains. VIII, 123 pages. 2003.

Vol. 527: C. E. Metz, Information Dissemination in Currency Crises. XI, 231 pages. 2003.

Vol. 528: R. Stolletz, Performance Analysis and Optimization of Inbound Call Centers. X, 219 pages. 2003.

Vol. 529: W. Krabs, S. W. Pickl, Analysis, Controllability and Optimization of Time-Discrete Systems and Dynamical Games. XII, 187 pages. 2003.

Vol. 530: R. Wapler, Unemployment, Market Structure and Growth. XXVII, 207 pages. 2003.

Vol. 531: M. Gallegati, A. Kirman, M. Marsili (Eds.), The Complex Dynamics of Economic Interaction. XV, 402 pages, 2004.

Vol. 532: K. Marti, Y. Ermoliev, G. Pflug (Eds.), Dynamic Stochastic Optimization. VIII, 336 pages. 2004.

Vol. 533: G. Dudek, Collaborative Planning in Supply Chains. X, 234 pages. 2004.

Vol. 534: M. Runkel, Environmental and Resource Policy for Consumer Durables. X, 197 pages. 2004.

Vol. 535: X. Gandibleux, M. Sevaux, K. Sörensen, V. T'kindt (Eds.), Metaheuristics for Multiobjective Optimisation. IX, 249 pages. 2004.

Vol. 536: R. Brüggemann, Model Reduction Methods for Vector Autoregressive Processes. X, 218 pages. 2004.

Vol. 537: A. Esser, Pricing in (In)Complete Markets. XI, 122 pages, 2004.

Vol. 538: S. Kokot, The Econometrics of Sequential Trade Models. XI, 193 pages. 2004.

Vol. 539: N. Hautsch, Modelling Irregularly Spaced Financial Data. XII, 291 pages. 2004.

Vol. 540: H. Kraft, Optimal Portfolios with Stochastic Interest Rates and Defaultable Assets. X, 173 pages. 2004.

Vol. 541: G.-y. Chen, X. Huang, X. Yang, Vector Optimization. X, 306 pages. 2005.

Vol. 542: J. Lingens, Union Wage Bargaining and Economic Growth. XIII, 199 pages. 2004.

Vol. 543: C. Benkert, Default Risk in Bond and Credit Derivatives Markets. IX, 135 pages. 2004.

Vol. 544: B. Fleischmann, A. Klose, Distribution Logistics. X, 284 pages. 2004.

Vol. 545: R. Hafner, Stochastic Implied Volatility. XI, 229 pages. 2004.

Vol. 546: D. Quadt, Lot-Sizing and Scheduling for Flexible Flow Lines. XVIII, 227 pages. 2004.

Vol. 547: M. Wildi, Signal Extraction. XI, 279 pages. 2005.

Vol. 548: D. Kuhn, Generalized Bounds for Convex Multistage Stochastic Programs. XI, 190 pages. 2005.

Vol. 549: G. N. Krieg, Kanban-Controlled Manufacturing Systems. IX, 236 pages. 2005.

Vol. 550: T. Lux, S. Reitz, E. Samanidou, Nonlinear Dynamics and Heterogeneous Interacting Agents. XIII, 327 pages. 2005.

Vol. 551: J. Leskow, M. Puchet Anyul, L. F. Punzo, New Tools of Economic Dynamics. XIX, 392 pages. 2005.

Vol. 552: C. Suerie, Time Continuity in Discrete Time Models. XVIII, 229 pages. 2005.

Vol. 553: B. Mönch, Strategic Trading in Illiquid Markets. XIII, 116 pages. 2005.

Vol. 554: R. Foellmi, Consumption Structure and Macroeconomics. IX, 152 pages. 2005.

Vol. 555: J. Wenzelburger, Learning in Economic Systems with Expectations Feedback (planned) 2005.

Vol. 556: R. Branzei, D. Dimitrov, S. Tijs, Models in Cooperative Game Theory. VIII, 135 pages. 2005.

Vol. 557: S. Barbaro, Equity and Efficiency Considerations of Public Higer Education. XII, 128 pages. 2005.

Vol. 558: M. Faliva, M. G. Zoia, Topics in Dynamic Model Analysis. X, 144 pages. 2005.

Vol. 559: M. Schulmerich, Real Options Valuation. XVI, 357 pages. 2005.

Vol. 560: A. von Schemde, Index and Stability in Bimatrix Games. X, 151 pages. 2005.

Vol. 561: H. Bobzin, Principles of Network Economics. XX, 390 pages. 2006. (planned)

Vol. 562: T. Langenberg, Standardization and Expectations. IX, 122 pages. 2006. (planned)

Vol. 563: A. Seeger, Recent Advances in Optimization. XV, 455 pages. 2006. (planned)

Vol. 564: P. Mathieu, B. Beaufils, O. Brandouy, (Eds.) Artificial Economics. XIII, 237 pages. 2005.

Vol. 565: W. Lemke, Term Structure Modeling and Estimation in a State Space Framework. IX, 224 pages. 2006.

Vol. 566: M. Genser, A Structural Framework for the Pricing of Corporate Securities. XIX, 185 pages. 2006.

GPSR Compliance

The European Union's (EU) General Product Safety Regulation (GPSR) is a set of rules that requires consumer products to be safe and our obligations to ensure this.

If you have any concerns about our products, you can contact us on ProductSafety@springernature.com

In case Publisher is established outside the EU, the EU authorized representative is:

Springer Nature Customer Service Center GmbH
Europaplatz 3
69115 Heidelberg, Germany

Batch number: 10844675

Printed by Printforce, the Netherlands